Faces in the Brain - a Behavioral, Eye-tracking and High-level Adaptation Approach to Human Face Perception

Dissertation

zur Erlangung des Grades eines Doktors
der Naturwissenschaften

der Fakultät für Biologie
und
der Medizinischen Fakultät
der Eberhard-Karls-Universität Tübingen

vorgelegt
von

Regine G. M. Armann

aus Würzburg

Dezember 2010

Bibliographic information published by the Deutsche Nationalbibliothek

The Deutsche Nationalbibliothek lists this publication in the Deutsche Nationalbibliografie; detailed bibliographic data are available in the Internet at http://dnb.d-nb.de .

ISBN 978-3-8325-2900-0

Logos Verlag Berlin GmbH
Comeniushof, Gubener Str. 47,
10243 Berlin
Tel.: +49 (0)30 42 85 10 90
Fax: +49 (0)30 42 85 10 92
INTERNET: http://www.logos-verlag.de

Tag der mündlichen Prüfung:	25. Februar 2011
Dekan der Fakultät für Biologie:	Prof. Dr. F. Schöffl
Dekan der Medizinischen	Prof. Dr. I. B. Autenrieth
1. Berichterstatter:	Prof. Dr. Heinrich H. Bülthoff
2. Berichterstatter:	Prof. Dr. Christian Wallraven

Prüfungskommission:

Prof. Dr. Heinrich H. Bülthoff

Prof. Dr. Christian Wallraven

Dr. Quoc Vuong

Prof. Dr. Hanspeter Mallot

Dr. Isabelle Bülthoff

I hereby declare that I have produced the work entitled: *"Faces in the Brain - a Behavioral, Eye-tracking and High-level Adaptation Approach to Human Face Perception"*, submitted for the award of a doctorate, on my own (without external help), have used only the sources and aids indicated and have marked passages included from other works, whether verbatim or in content, as such. I swear upon oath that these statements are true and that I have not concealed anything. I am aware that making a false declaration under oath is punishable by a term of imprisonment of up to three years or by a fine.

Regine Armann

Acknowledgements

I am indebted to a number of people who helped and supported me during the stages of this dissertation.

First, I would like to thank Prof. H.H. Bülthoff for giving me the opportunity to work in his department at the Max Planck Institute for Biological Cybernetics in Tübingen, which provides an inspiring and challenging environment for a student to gain knowledge and accomplish projects that finally lead to a PhD.

Second, but in fact foremost, I want to express my special gratitude to Dr. Isabelle Bülthoff, for accepting me for my diploma thesis and later on as a PhD student, for her enduring supervision and support and enthusiasm, for initiating me into the world of and a life in science, for her patience in tough times and innumerous but nevertheless very creative comments, thoughts and revisions of my work.

I would also like to acknowledge Prof. Martin Giese, Dr. Uta Noppeney and Prof. Christian Wallraven, as members of my advisory board, for comments and fruitful discussions of my work. Prof. Gillian Rhodes and Dr. Linda Jeffery from the University of Western Australia I want to thank for a fantastic and very instructive stay in their lab in Perth and for their collaboration, together with Andrew Calder from the MRC Cognition and Brain Sciences Unit in Cambridge, UK, on the project presented in Chapter 3.

Johannes Schultz I owe special thanks for getting me started and continuously support me on Matlab programming and fMRI, and for thousands of insightful discussions, stimulating brainstorming sessions, day-to-day problem solutions, and a lot of patience during the various stages of this thesis.

Many thanks also to the other members of the Recognition & Categorization group, former and currents ones, for helpful ideas, constructive criticism, interesting conversations, scientific and less scientific ones, for making work fun and to everybody who helped me to endure and not give up during the final stages of this thesis.

Thanks also to the administrative staff at the Max Planck Institute and the Graduate School, especially Dagmar Maier and Katja Deiss, and for the system administrators, who really help to make life as easy as possible and to concentrate on the science part of it.

Finally, I am very grateful to my parents for their continuous support during my studies, my diploma thesis and eventually this dissertation; without them, some of the things I did would not have been possible and especially the last stages of my thesis would have been much more difficult.

Tübingen, Dezember 2010 Regine Armann

Table of Contents

Synopsis

Gesichter zu erkennen ist eine der bemerkenswertesten Fähigkeiten des menschlichen Gehirns. Ein Gesicht verrät uns nicht nur das Alter, das Geschlecht, die Intention und die Gemütslage einer Person, darüberhinaus können wir auch eine scheinbar unendliche Zahl an Individuen anhand von eher subtilen Unterschieden in ihren Gesichtszügen unterscheiden und identifizieren. Wie das Gehirn alle diese Aufgaben löst, das heisst, wie Gesichter betrachtet, beurteilt, unterschieden und wiedererkannt werden, ist eine sehr alte Fragestellung im Bereich der visuellen Forschung. Die Arbeit, die in dieser Dissertation vorgestellt wird, soll dazu beitragen, manche Aspekte dieser Frage genauer zu beleuchten. Drei Studien wurden durchgeführt, mit drei unterschiedlichen Ansätzen und Methoden. Eine Studie, in der die Augenbewegungen von Probanden aufgezeichnet wurden, beschäftigt sich mit der Frage, welche Information in einem Gesicht als relevant für verschiedene Aufgaben angesehen wird und was uns das über die Repräsentation von Gesichtern im Gehirn verrät. In einer psychophysischen Studie soll geklärt werden, ob männliche und weibliche Gesichter als zwei unterschiedliche Kategorien wahrgenommen werden und ob Information über das Geschlecht und die Identität eines Gesichts tatsächlich, wie lange angenommen, unabhängig voneinander verarbeitet wird. Wie genau Gesichter unterschiedlicher Ethnien im Gehirn repräsentiert werden und wie demzufolge der sogenannte „Face Space" aussehen könnte, wird in einer dritten Studie untersucht, mithilfe einer Adaptationsmethode bei der, analog beispielsweise zum Farbensehen, durch visuelle Exposition Nacheffekte erzeugt werden. Die Ergebnisse dieser Studien sollen einen Beitrag zum besseren Verständnis des Wahrnehmens und der Verarbeitung menschlicher Gesichter leisten, und damit, angesichts der sozialen und biologischen Relevanz dieser grundlegenden Fähigkeit für den Menschen, einen Beitrag zum generellen Verständnis perzeptueller und kognitiver Prozesse im Gehirn.

Introduction

Recognizing faces is a one of the most common tasks we perform in everyday life. How good we manage to remember and identify other people, or whether we have a deficit in this skill, has important implications for effectively interacting in social environments. Generally, we seem to deal with the task of identifying and discriminating innumerable face identities in a very accurate and effortless way. This is impressive, considering that faces are very similar to each other, more than most object categories that we come across in everyday life. Faces share all the same basic configuration: a nose placed in the center between two eyes above and a mouth below. Differences between two identities are therefore rather subtle, compared to the differences between several chairs or cars. How the brain copes with this demand and more precisely how faces are perceived, processed, encoded and represented in the brain, is the topic of this thesis.

We can recognize individuals on the basis of their hair style, voice or even their way of moving, but looking at a face is the most reliable way to identify a person. Besides, a face, even that of a person we do not know, contains a lot more information: We see in it whether somebody is male or female, how old he or she is, how that person is feeling or what their intention towards us might be; we find them attractive or not, and after a brief look a somebody's face we even have an opinion on whether that person is intelligent or trustworthy. It is therefore not surprising that human faces, as a class of visual objects, have probably undergone more investigation than any other object class. Evidence regarding different aspects of face perception and processing comes from a variety of fields and approaches within the broad domain of vision research. Studies date back to the 1950s and include psychological and psychophysical methods, eye-tracking, computer vision, brain imaging and other techniques.

Several lines of research have suggested that the perception of faces, as compared to recognition of other objects, may be "special". Faces are preferentially attended by young infants compared to other complex visual stimuli or even scrambled faces (e.g., Bruce, 1986). Yarbus (1967) showed that faces preferentially draw the attention of adult viewers when placed in complex scenes. Moreover, the neural systems underlying face perception seem to be partially independent of those systems that underlie object perception in general (Haxby et al., 2000), and even in newborns, neural substrates involved in face perception have been localized (Farah et al., 2000). Neurophysiological studies have demonstrated that face recognition can be selectively impaired when these areas are damaged, a phenomenon called "prosopagnosia", while recognition of objects of equivalent complexity can remain intact (Farah, Klein, & Levinson, 1995; but see for example Tarr & Gauthier, 2000, for a divergent opinion). These findings imply that people use different brain areas for face recognition than for the recognition of other types of objects. Furthermore, single cell recordings and brain imaging in the temporal cortex of monkeys (Desimone, 1991; Tsao et al., 2001; 2008) showing that populations of cells respond selectively to faces also indicate faces are "special" in that they are processed differently from other object classes.

Brain imaging studies (e.g. Kanwisher et al, 1997; Grill-Spector et al., 2004) typically show enhanced activity when observers are being presented with faces primarily (but not only, see e.g., Haxby et al., 2000; Tsao, Moeller, & Freiwald, 2008) in an area of the temporal lobe known as the fusiform gyrus, often referred to as the fusiform face area or FFA. However, it has been suggested that this area may be more generally devoted to the recognition of exemplars of a well-learned object class (Gauthier et al, 1999, Tarr & Gauthier, 2000) and that the expertise that humans have with faces simply exceeds the expertise in any other commonly encountered object category.

In brief, whether specific brain areas and skills are devoted exclusively to recognizing faces or whether face recognition is just part of a general skill for making fine within-category discriminations is still a matter of active debate. There is no doubt, however, that face perception is a fundamental, biologically and socially relevant skill of the human visual system. Understanding more about the perceptual, cognitive and neural mechanisms underlying human face processing expertise shall therefore contribute to a better comprehension of processing in the human brain in general.

The field of face perception research is an extremely broad one, including a high number of different subareas engaged in very specific questions and using all sorts of different approaches and methods (to illustrate that: a search for simply "face" in *ISI Web of Knowledge*[SM] exceeds the maximum number of 100.000 results to be displayed; searching for "face recognition" still yields nearly 36.000 results). It would therefore be impossible to provide a comprehensive overview that would only remotely do justice to the scope and complexity of the subject. In this chapter, I will hence simply depict some key aspects of human face perception, and introduce the concepts underlying the experimental work that went into this dissertation. I will then set out the aim and structure of the thesis and outline the studies that have been carried out.

Frameworks for face processing

More than twenty years ago, Vicky Bruce and Andy Young presented a theoretical model of how familiar faces are recognized (Bruce & Young, 1986). Their model argues that recognition involves several components gathering distinct types of information from faces: The processing of simple physical aspects of the face leads to a structural representation containing information about for example the age, sex, or attractiveness of a person, and from so-called "person identity nodes", semantic information such as the corresponding name can be derived. Recognition of familiar faces involves a match between the result of structural encoding and previously stored information, held in "face recognition units". One crucial point of this framework is that it proposes a separation between the processing of several characteristics of faces, e.g., between the processing of identity and the processing of more "changeable aspects" like facial expressions. At the time, the model of Bruce and Young was supported by evidence drawn from laboratory experiments and studies of patients with different types of cerebral injury. Although imprecise about how exactly individual faces are represented and distinguished, and vague at the level of neural implementation, this cognitive framework has remained the dominant account of face perception until today and has been very influential for the understanding of face recognition at a functional level. Only recently, in light of new evidence and especially from neurophysiological data, there have been some attempts to offer an alternative account of human face recognition (e.g., Haxby et al., 2000; Calder & Young, 2005).

The face space idea

A framework for face recognition can also approach faces from a stimulus point of view, by asking how the brain could most efficiently represent the characteristics of faces and the similarities or differences between them. Such a framework was proposed by Valentine (1991). The key idea of this model is that faces can be represented as points or vectors in a multi-dimensional space, the dimensions of this space representing the properties in which faces differ from each other (e.g., nose length, face width, etc.). The coordinates of a face in this space are defined by its values on each dimension. As a simplified example, imagine a two-dimensional space in which all faces differ in only two features, for example the roundness of the face outline and the color of the face (see *Figure 1*). The coordinates for each point in this space would determine the color and the roundness for each individual face. By definition, the center of the face space corresponds to the average face of the represented population. The closer a face lies to the average in the space the more "typical" it is, while faces lying further away from the center are perceived as looking more distinctive. Also, the closer two faces are in the space, the more similar they are to each other. Valentine assumed that the values of the feature dimensions of the population of faces experienced will vary normally around the central tendency. Thus the density of points (i.e. the number of previously seen faces) will decrease as the distance from the central tendency increases (but see for example Burton & Vokey (1998) for an opinion on why this assumption is too unspecific – without, however, generally undermining the explanations that Valentine's framework offers). This framework is suitable to explain numerous findings in human face perception, e.g., the recognition advantage for distinctive versus typical faces (Valentine, 1991) or the phenomenon that faces of one's own race are more easily recognized and discriminated from each other than faces of an unfamiliar race (the so-called "other race effect"). There is also recent fMRI data providing evidence for a neural face space representation in the brain (Loffler et al., 2005): In this study, neural activity was recorded from a face responsive area in the fusiform gyrus (FFA) while participants observed synthetic face identities with varying facial geometry (head shape, hair line, internal feature size and placement). fMRI activation in the FFA of observers was found to increase as a function of increasing distance of face characteristics from the average face. Leopold, Bondar and Giese (2006) recorded from single neurons in the monkey inferotemporal cortex, while monkeys were presented with a large number of faces corresponding to different positions in a computational face space. Their data suggests that neurons are tuned relative to the average face, as firing rates increased with increasing distance of face stimuli from the average.

What this framework remains silent about is the nature and the number of dimensions necessary to represent faces most efficiently. There are obviously very simple dimensions that serve to distinguish between faces, like the size, shape or color of specific features. Some dimensions, however, must be more „holistic“ (in the sense that they include the whole face), for example the sex or race of a face, or even more abstract than that: "Attractiveness" or even "likeability" might be such characteristics that we can use to classify faces although it would be hard to specifically state what features or feature combination make a face likeable.

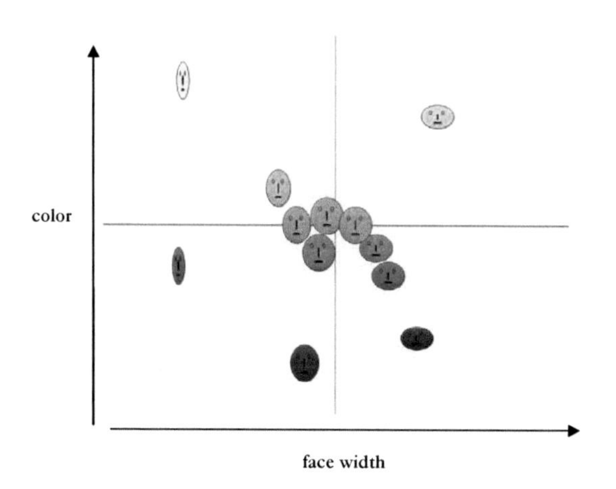

Figure 1. A simplified two-dimensional version of the multi-dimensional face space framework as proposed by Valentine (1991).

Norms for the encoding of faces

Another crucial point that is not specified in Valentine's framework is the way how exactly faces are encoded within the multi-dimensional space. Valentine proposed two slightly different versions of such encoding. (1) A "norm-based" model in which faces are represented as deviations (vectors) from an abstracted population mean located at the origin of the space, i.e., representing the central tendency. (2) A purely "exemplar-based" model that represents each face as a point in space, in terms of its distance to other faces. Note that in the latter model, the origin of the space plays no part in encoding faces; it merely indicates the point of maximum exemplar density. Since both models make similar predictions concerning a number of different effects in face perception (e.g., the effects of distinctiveness and race, see above), albeit in different ways, Valentine did not distinguish between the two.

There has been considerable debate over the two models (e.g., Byatt & Rhodes, 1998; Rhodes, Brennan & Carey, 1987; Rhodes, Carey, Byatt & Profitt, 1998; Valentine, 1991; Valentine & Endo, 1992), but recently there is growing evidence in favor of norm-based rather than exemplar-based coding. Behavioral studies show that information about the schema of a given set of patterns is abstracted from stored instances with very high efficiency; e.g., after learning a set of dot patterns, the prototype of that set is more easily classified than control patterns (Posner & Keele, 1968). As to the perception of face stimuli, it has repeatedly been found that observers spontaneously abstract averages from a set of faces they have seen before (Bruce et al, 1991; Walton & Bower, 1993; Inn et al., 1993). The ease with which we can identify caricatures, which means exaggerating how a face deviates from the norm (increasing its distance from the norm along its unique identity vector) also seems to support some form of norm-based coding (Benson & Perrett, 1994; Byatt & Rhodes, 1998; Rhodes et al., 1987, 1998; Calder, Young, Benson & Perrett, 1996; Lee et al., 2000). As described above, neural data have shown that with increasing distance from the average, faces elicit stronger fMRI activation in the human FFA (Loffler et al., 2005); and they also increase firing rates of face-selective neurons in monkey anterior inferotemporal cortex (Leopold, Bondar, & Giese, 2006).

More evidence that faces are coded relative to a norm in face space has emerged from high-level face adaptation methods. Adaptation or prolonged exposure to a stimulus can cause temporary changes (aftereffects) in perception. A familiar example of a low-level aftereffect is the experience of seeing still objects move upwards after staring at a waterfall for several seconds (Mather et al., 1998). Neurologically, adaptation leads to a change in the average tuning of a population of neurons (Barlow & Hill, 1963), by causing a temporary reduction in activity from neurons responding to the adapted stimulus. As a consequence, a previously neutral stimulus appears altered in the opposite way, with respect to the adapting stimulus (see Clifford and Rhodes, 2005, for a recent review on neural coding mechanisms and how they illustrate the dynamic nature of visual representations). Perceptual aftereffects resulting from short-term adaptation (seconds to minutes) have been widely used to explore the principles underlying perceptual coding, leading them to be dubbed the "psychologist's microelectrode" (Frisby 1980). Aftereffects have provided insight into coding properties of color, orientation, curvature and a number of other basic visual stimulus attributes. In the last decade, such aftereffects have also been reported for more high-level stimuli like human faces. Prolonged exposure to face stimuli can result in the specific misperception of numerous facial attributes, including sex, race, expression, normality, attractiveness and eye gaze direction (Hurlburt 2000; Jenkins et al., 2006; O'Leary & McMahon 1991; MacLin & Webster, 2001; Rutherford et al., 2008; Watson & Clifford, 2003; Webster & MacLin, 1999; Webster et al., 2004).

Webster and colleagues, for example, adapted observers to faces of different sexes (male-female), races (Asian-Caucasian), or expressions (happy-sad). When testing observers later on morphs that were created in between the endpoints of these categories, they found that the perception of what appeared to be a "normal" member of a category along the continuum was consistently shifted towards the adapted category. For example, staring at a male face for several seconds makes an ambiguous face look female – apparently, the perceptual (ambiguous) midpoint has been shifted towards the male end of the continuum. Leopold, O'Toole, Vetter and Blanz (2001) reported a study in which they created face pairs that fell on opposite sides of the average in a computational face space (*see Figure 2*). The face pairs had thus opposite features, e.g., large eyes versus small eyes, or thin face versus round face. A face and its "antiface" that was created by morphing through the average of the space appeared like two completely distinct and unrelated identities (see *Figure 2* for examples of such face – antiface pairs). The authors found that adapting participants to an anti-face shifts the average towards the adapted face, so that it takes on the identity of the respective opposite face identity. Despite these faces appearing unrelated to one another, the findings indicate that they were perceptually linked via their relationship to the average face, suggesting that it functions as a norm relative to which identity is coded.

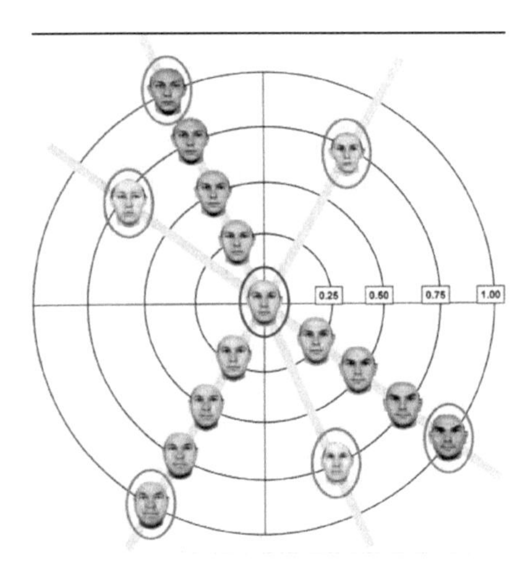

Figure 2: Computational face space created from three original identities (green circles), and their respective anti-faces (orange circles), with reference to the average face of the space (blue circle). From: Leopold, O'Toole, Vetter & Blanz, 2001.

Facial features and holistic processing

As described so far, there is a great deal of evidence for a norm-based face representation in the brain. However, the fact that the potential dimensions that are used to code faces in such a model are still a matter of ongoing investigation raises the question which information in a face stimulus is really gathered and processed by an observer.

A face consists of specific features whose characteristics can be processed independently from each other, e.g., the shape of nose and mouth, and the color of the eyes. These features also correlate with each other in a spatial arrangement, and there is good evidence that faces are processed as "configurations" rather than only mere sets of individual elements (e.g., Young, Hellawell & Hay, 1987). Parts of a familiar face (e.g., mouth, eyes) are more easily recognized when presented in the context of the whole face than when presented as isolated parts (Tanaka & Farah, 1993). For parts and wholes of scrambled faces, inverted faces and houses, the recognition difference between presentation in isolation and presentation as a whole is much smaller. Another very clear demonstration of holistic processing comes from studies with chimeric faces in which the top half of one face is aligned with the bottom half of another. When participants are instructed to identify the top half of a composite face, they are impaired, presumably because holistic processing prevents them from attending exclusively to the features in the top half and ignoring the whole Gestalt (Carey and Diamond 1994; Hole 1994; Young et al 1987). Note, however, that there are different ways of defining "holistic" processing, varying primarily in the role they attribute to the processing of facial features. In the original concept by Tanaka & Farah (1993), parts of a face were not explicitly represented at all. On the basis of data showing that featural information alone can be sufficient for recognizing faces and that features are also processed and stored independently of the configuration of the face, Schwaninger and colleagues (e.g., Schwaninger et al., 2009) proposed a "dual-code" model. Here, featural and configural information is represented separately before it converges into a face representation that is indeed "holistic".

The importance of the overall configuration of a face can help us understand why face recognition and in particular face identification can be remarkably robust to a variety of unnatural transformations (see e.g., recognition experiments using faces with low spatial frequency information or spatial distortions: Sinha et al., 2006) as well as natural transformations (for example expressions). However, since information about the parts of a face and their configuration interact (Tanaka & Sengco, 1997), for example, manipulating only the eyes or the mouth usually also leads to a change in the configuration of the whole face. Besides, facial features are not themselves all equally important for recognition. The external features of the face (hairstyle and face outline) seem to be more important in the recognition of unfamiliar faces, while recognition of familiar faces seems to rely more heavily on internal features (Shepherd, Davies & Ellis, 1979) or even on eyes or eyebrows alone (O'Donnell & Bruce, 2001; Sinha, et al., 2006).

Of the internal features, the eye region but not the chin seems to be the most important for learning and identification of faces (Falk et al., 2000).

The majority of studies investigating different types of information present in faces use psychophysical methods and paradigms, where participants for example judge, discriminate, recognize or compare face stimuli on a screen (e.g., Hill, Bruce & Akamatsu, 1995; O'Donnell & Bruce, 2001; Goffaux et al., 2005; Boutet, Collin & Faubert, 2003). In such studies, stimulus manipulations and subsequent measures of observers' performance are used to assess which changes in facial information observers are sensitive to. Recently, the number of studies using eye-tracking approaches in face perception tasks is growing (e.g., Barton et al., 2006; Galpin & Underwood, 2005; Henderson et al., 2003; Stacey et al., 2005; Hsiao & Cottrell, 2008), thereby taking into consideration that visual acuity varies across the visual field, and that only the part of the stimulus close to the small foveal regions surrounding fixation is perceived with high acuity (e.g., Anstis, 1974). Thus, when an observer is viewing any picture or scene, he will see only a small part of the stimulus in foveal vision and need to constantly scan the image to gather all relevant information. Records of eye movements (Yarbus, 1967; Loftus et al, 1978) have shown that foveal vision is indeed not randomly distributed across stimuli, but mainly directed to visual elements containing information essential to the observer (see *Figure 3* for examples of eye movements on face stimuli). Henderson and colleagues found recently that eye movements play a functional role in learning human faces and that restricting the eye gaze of observers to the center of a face during a learning phase clearly impairs later recognition performance (Henderson, Williams, & Falk, 2005).

In brief, following observers' focus of attention over time during face perception tasks gives us not only information about where somebody looks on a face, it also reveals the information that the observer (consciously or not, considering that not all eye-movements are voluntary) considers relevant. Both of these types of information provide valuable evidence on cognitive strategies and processes involved in the perception of human faces.

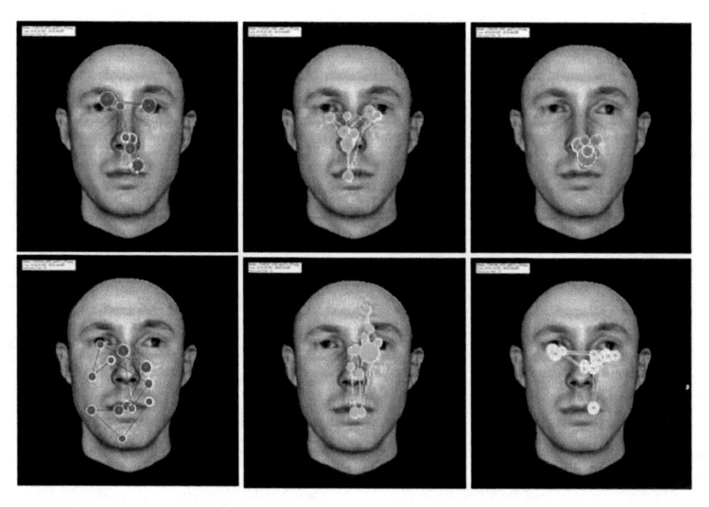

Figure 3: Exemplars of eye-movement data on human face stimuli, recorded with a Tobii T60 XL eye-tracker during a race classification task. Each color represents data from a different observer. As is apparent from the recordings, most fixations are directed to the eyes, the nose and the mouth of the face image, suggesting that these features are relevant to accomplish the task at hand. Some observers, however, scan only the center of the face (e.g., orange recording left upper row), while others distribute more fixations all over the internal face region (see dark blue recording left lower row).

The Morphable Model

Most face perception experiments are done using photographs of faces as stimuli, either in their original form or manually altered (e.g., Ellis, Burton, Young & Flude, 1997; Tanaka & Sengco, 1997). Although this allows staying very close to the natural appearance of faces, these stimuli have a number of important limitations. Faces derived from photographs are rarely matched (and thus not controlled) for size, orientation or lighting conditions. Neither do photographs allow modifying face specific image properties in a systematic way. Similarities between such stimuli can thus be measured, controlled or manipulated only up to a certain extent. Without direct control over the actual stimuli used in experiments, it is sometimes difficult to empirically test an assumption or to reliably draw conclusions from an experiment.

Recently, more simplified face stimuli have been used, to overcome some of these constraints. Wilson, Loffler & Wilkinson (2002), for example, created a set of synthetic faces that are computerized line drawings obtained from gray-scale face photographs.

Here, landmark points within each face are manually identified and local contrast information in specified regions is extracted. By interpolating the landmark points and placing face features in specified locations, synthetic faces are reconstructed from the photographs. Wilson and colleagues could show that observers can accurately match such synthetic faces to original photographs across various viewpoints. The stimuli also produce inversion effects, as reported for face photographs (e.g., Yin 1969; Valentine 1991). A criticism of this method is, however, that each face is reconstructed on the basis of generic face features (such as eyes, nose, mouth) that have to be predefined. Recent studies suggest that for characterizing the perceptually salient information available in faces, it might be more useful to derive "features" empirically (e.g., Schyns, Bonnar & Gosselin, 2002).

Another very recent method, "silhouetted face profiles" (Davidenko, 2007) shares some properties with the synthetic face stimuli, while using a purely shape-based approach to face representation. The stimuli are black-and-white images derived from gray-scale photographs of face profiles, cropped at the forehead, below the chin, and behind the ear line. Observers are able to reliably extract sex and age from the silhouettes, and their judgments of attractiveness made on the silhouettes correspond to those judgments on profile- and front-view images of the same faces (Davidenko, 2007). An inversion effect has also been found using these silhouette stimuli, suggesting that they are processed similarly to normal face stimuli.

There are, however, some clear limitations to the face silhouette approach. Certain features of a face, like the eyes, are particularly important for face perception (e.g., Brown & Perret, 1993; Bruce et al., 1993) – and these parts are not visible in silhouettes. Neither can texture and color information be retrieved from them, which is also known to play an important role in recognition and the perception of sex, age, race and other face characteristics (e.g., Alley & Schultheis, 2001; O'Toole, Vetter & Blanz, 1999; Yip & Sinha, 2002).

The method used to create the stimuli for the studies presented in this thesis is based on 3D laser scans of real heads (Blanz & Vetter, 1999). Here, several thousand locations and textures across a collection of 3D face data are automatically aligned and specialized graphics software is used to display and manipulate the resulting images. Starting from an example set of 3D face models, a "morphable model" is derived by transforming the shape and texture of the examples into a vector space representation. New faces can be modeled by forming linear combinations of the prototypes, and manipulations can be done according to complex parameters such as sex or distinctiveness. Shape and texture constraints derived from the statistics of the example faces are used to regulate the "naturalness" of the modeled faces and avoid faces with an unlikely appearance. Importantly, variability across faces along various dimensions is used to create and manipulate stimuli, but without specifying a priori what those dimensions or features are. Such a "featureless" representation enables us to study what the perceptually salient aspects of a face are, without presupposing classical or intuitive features.

The Max Planck 3D head database (http://faces.kyb.tuebingen.mpg.de) consists of a collection of more than 200 laser scans of real heads. As described above, the heads are stored as vectors in terms of their distance from a reference head (Blanz, 2000); the database therefore represents a kind of "face space" in itself. By combining it with the morphing technique developed by Blanz and Vetter (A Morphable Model For The Synthesis Of 3D Faces, 1999), we can manipulate single properties of faces (e.g., only their sex), while leaving other characteristics of the face stimuli constant.

Aim and structure of the thesis

The aim of this thesis is to explore different aspects of face perception and thereby contribute to the growing evidence on how faces are processed, encoded, stored and retrieved from memory. The studies that have been conducted in the course of this dissertation pertain to different approaches and research questions. This thesis is therefore structured into three main chapters presenting the three main studies.

In **Chapter 1**, information processing during face tasks was explored using eye tracking methodology. Studies have shown that featural information that is considered to be relevant is looked at preferentially (see above), thus eye tracking methodology provides us with an objective insight into cognitive processes involved in face perception. Eye movements of human participants were recorded while they compared two faces presented simultaneously. Observers' viewing behavior and performance were examined in different tasks, and the facial characteristics that were manipulated were the identity and the sex of the face stimuli. Frequency, duration, and temporal sequence of fixations on previously defined areas of interest in the faces were analyzed. The results corroborate an earlier finding (e.g., Schyns, Bonnar, & Gosselin, 2002; Pearson et al., 2003), that is, where observers look in a face depends on the task they are performing; they moreover expand this finding into the realm of side by side face comparison. Interestingly, the data also reveal scanning asymmetries that have not been reported before in the literature, suggesting that in face comparison task, participants (wrongly) assume that faces are generally symmetrical. Moreover, the scanning patterns show very specific differences in male and female participants' viewing behavior, suggesting that both groups of observers have different representations of how male and female faces differ from each other.

Chapter 2 presents a study that goes further into the question how the sex of faces is represented in the brain. To find out whether male and female faces are represented as two distinct categories, here, we examined categorical perception (CP) of the sex of unfamiliar face identities. Categorical perception is a fundamental cognitive process that enables us to sort similar objects in the world into meaningful categories with clear boundaries between them. Objects within a category are perceived as more similar to each other than to objects belonging to another category even if the physical differences between them are equal. The most well-known example of CP is our perception of a rainbow, which constitutes a monotonic increase of wavelength of visible light – still, we see a series of discrete color bands. CP was first observed with auditory stimuli (e.g., Burns and Ward 1978; Liberman et al. 1967) and color (Bornstein and Korda 1984), but it has since been found in a variety of domains. Recently, CP has been found for high-level stimuli like human faces, more precisely, for the perception of face identity, expression and race. For sex however, which represents another important and biologically relevant dimension of human faces, results have been equivocal. Here, we reinvestigate CP for sex using newly created face stimuli to control two factors (the degree of femininity or masculinity, and the degree of familiarity with the face identities) that might have influenced the results in earlier studies. Our findings speak against the classical Bruce & Young model (see above) which proposed that the different sub-functions of face processing are computed independently. At the core of this architecture is the distinction between an individual face recognition pathway, aimed at identifying the person, and parallel pathways involved in processing facial expression, facial speech, and "visually derived semantic information" from the face such as sex, age, or race (Bruce & Young,

1986). Our findings rather propose interactive processing of information related to the sex and the identity of a person.

The study reported in **Chapter 3** directly addresses the question whether the multi-dimensional face space framework described above is implemented in the brain and how exactly faces might be coded within such a space. Faces share many visual properties and could be encoded in one face space against one single norm. However, certain face properties like sex, race or age may result in grouping of similar faces each with their own norm face. How faces might be "sub-classified" in face space remains thus to be determined. Faces of different races represent distinct visual and social categories, and faces of one's own race are usually more easily recognized than faces of a more unfamiliar race (e.g., Meissner & Brigham, 2001). Current face space models therefore propose that own- and other-race faces are represented in different locations of face space (Valentine, 1991; Valentine & Endo, 1992), with the other-race faces being grouped far away from the center of the space. The aim of this study was thus to find out whether faces of different race categories are coded against their own respective norm, or against a generic common norm for all faces in faces space. To that end, a high-level adaptation paradigm was used, where exposure to an adaptor face systematically distorts the perception of a subsequently viewed test face towards the "opposite" identity in face space (see *Figure 2* above). We measured identity aftereffects for pairs of adaptor-test faces that were created as lying opposite race-specific (Asian and Caucasian) averages and pairs that were opposite a "generic" average (both races mixed together) in a computational face space. The results suggest that Asian and Caucasian faces are coded using race-specific norms. Moreover, they indicate that norms, as has long been proposed but rarely supported by evidence, also have a functional role in face processing.

Finally, the implications of the findings presented here and possible avenues for future research are discussed in **Chapter 4**.

Declaration of contribution of the candidate

This thesis is presented in the form of a collection of manuscripts that are, at the time of thesis submission, either published or prepared for publication. The bibliographic details of the studies and where they appear in the thesis are set out below, together with a description of the contribution of each author.

Chapter 1

Armann, R., Bülthoff, I. (2009). Gaze behavior in face comparison: The roles of sex, task, and symmetry. *Attention, Perception, & Psychophysics* 71(5), 1107-1126.

Design, stimulus generation, experimental work and analysis of this study have predominantly and independently been developed and finalized by the candidate. The work has also been presented at several scientific conferences by the candidate and the present manuscript has entirely been written by the candidate. The co-author's role was that of a supervisor in giving advice, offering knowledge and criticism, and revising the manuscript.

Chapter 2

Armann, R., & Bülthoff, I. Male and female faces are only perceived categorically when linked to familiar identities - and when in doubt, he is a male. *(in preparation).*

The idea for this study was proposed by the candidate. Design, stimulus generation, experimental work and analysis of this study have predominantly and independently been developed and finalized by the candidate. The work has also been presented at several scientific conferences by the candidate. The present manuscript has entirely been written by the candidate. The co-author's role was that of a supervisor in giving advice, offering knowledge and criticism, and revising the manuscript.

Chapter 3

Armann, R., Jeffery, L., Calder, A.J., & Rhodes, G. Race-specific norms for coding face identity and a functional role for norms (submitted to *Journal of Vision*).

This study was accomplished in the Lab of Professor Gillian Rhodes at the University of Western Australia. The idea was proposed by the candidate; the face stimuli were generated by the candidate and brought to the host lab. Design, experimental work and analysis of the data have predominantly and independently been finalized by the candidate. The work has also been presented at a scientific conference by the candidate and the manuscript was written entirely by the candidate. The senior author's role was that of a supervisor in giving advice, offering knowledge and criticism, and revising the manuscript. The second and third author were involved in discussions about the experimental design and in the revision of the manuscript before it was submitted to a scientific journal.

Chapter 1: Gaze Behavior in Face Comparison: The Roles of Sex, Task, and Symmetry

Introduction

Looking at a face is the most reliable way to identify an individual, but in addition to recognizing familiar faces, we can derive other important information from the face of a person we do not know. Faces are central in human interactions as they provide critical information about the age, sex, identity, mood, and intention of another person. Numerous psychophysical studies have investigated various aspects of the process of face perception (see for example Bruce & Young, 1986; Farah, Wilson, Drain, & Tanaka, 1998; for a review: Bruce & Young, 1998). Recently, eye-tracking studies have brought new insight in the way we encode and perceive faces (e.g., Williams, Senior, David, Loughland, & Gordon, 2001; Pearson, Henderson, Schyns, & Gosselin, 2003; Stacey, Walker, & Underwood, 2005).

Since the eyes see only a small part of a visual scene at high resolution at a time, they are constantly moving to scan interesting features with the fovea. Records of eye movements (Yarbus, 1967; Loftus & Mackworth, 1978) have shown that foveal vision is mainly directed to visual elements containing information essential to the observer during perception. Henderson and colleagues (Henderson, Williams, & Falk, 2005) showed recently that eye movements play an important functional role in face learning as restricting the eye gaze of observers to the center of a face during a learning phase clearly impairs later recognition performance.

Pearson and colleagues recorded observers' gaze in different face perception experiments (Pearson et al., 2003), and found that not all information in the face images was looked at with the same frequency and duration, and that especially the task observers were asked to perform influenced viewing behavior. When instructed to classify faces by their sex, observers looked longer at the eyes of the face images, compared to a face identification task where fixation time was more distributed all over the face images. In the same study, the mouth was more often and longer looked at when the mood of the face was to be specified. This task-dependency of eye movements when viewing faces is corroborated by studies using the 'Bubble Method' (e.g., Schyns, Bonnar, & Gosselin, 2002) which show that, for example, eyes alone are 'diagnostic' for sex decisions but not for face identification. Both studies point out the importance of task-dependent attention allocation as opposed to stimulus-driven saliency involved in face perception.

Barton and colleagues have suggested that internal representations resulting from familiarity with a face influence scanning patterns (Barton, Radcliffe, Cherkasova, Edelmann, & Intriligator, 2006), as their study shows eye movement sequences towards famous faces to be more idiosyncratic, i.e., more different from each other, than the more standardized scanning patterns towards unknown faces.

Furthermore, the two halves of a face have been found to be of unequal importance: Observers tend to base their responses predominantly on the information on the left in sex classification (e.g., Butler, Gilchrist, Burt, Perret, Jones, & Harvey, 2005) as well as in face memory tasks (Gilbert & Bakan, 1973, Mertens, Siegmund, & Grüsser, 1993), or the judgement of face expressions (e.g., Christman & Hackworth, 1993). In general, this perceptual bias towards the left hemiface is interpreted as a consequence of a right-hemisphere specialization for face processing (e.g., Barton et al., 2006; Kanwisher, McDermott, & Chun, 1997).

All these findings demonstrate that eye movements can tell us something about how faces are perceived and represented. Although observers are not always consciously aware of it, they allocate their attention (i.e., eye gaze) to specific features of a face (and not to others), or in a specific pattern. It is important to note that most of the eye-tracking studies on face perception cited so far investigated gaze behavior of observers during learning or recognition of single face images. In that case, however, the scanning pattern might depend on memory representations of known face identities or of general facial characteristics (e.g., expressions) that are compared to the face image depending on the paradigm (task). Hence, only one side of the comparison, i.e., the presented stimulus, is available to the experimenter and can be manipulated, whereas internal representations remain inaccessible. Yet, in everyday situations, one also happens to compare faces of people physically present or to match the face of a person to a photograph. When interacting with the visual world in such a way the stimuli being compared remain visually available to the observer and thus, in an experiment using a direct comparison task, both sides of the comparison can be manipulated and are accessible for analysis. Furthermore, in the studies mentioned above that are most relevant to our work, observers were highly familiar (Pearson et al, 2003; Schyns et al, 2002) with the faces they had to identify or classify by their sex, or at least they saw the same faces over and over again (Gosselin & Schyns, 2001). Whereas familiarity with the test face is needed in an identification task, it is not required to classify a face's sex; in that case familiarization was nevertheless done in these studies to ensure participants' same level of experience with all faces. The results of these studies might thus pertain more to the specifics of gaze behavior when inspecting known faces than to information about diagnostic features generally used when observers identify or classify faces. On this account, we investigated eye movements when observers looked at two unfamiliar faces presented side by side; in this setting there is no necessary reference to some internal memory representation. We assume that eye fixations to specific facial features indicate the importance given to these features for the task at hand while specific scanning patterns reflect different strategies of information retrieval (e.g., based on local features, based on relations between features, or based on whole faces, etc.).

Thus the first of the questions that motivated this study was whether, despite the differences between single face experiments and face pair comparisons that have been mentioned, and when using strictly unfamiliar test faces, our participants would consider the same key facial features diagnostic as those reported in earlier studies. If this is the case, our results will not only confirm but extend the validity of these earlier findings to a general level.

Second, we asked how observers compared the two faces of a pair – they could compare each facial feature with the corresponding feature on the other face, analogous to the object-by-object strategy that has been found in comparative visual search tasks (Galpin & Underwood, 2005), or they could rather compare the faces as a 'whole', indicated for example by more extensive scanning patterns on each face before comparison with the other face, and/or relatively few comparison saccades between faces at all. Based on the repeatedly shown left hemiface bias in face perception tasks, we would also assume that observers compare rather the left half faces of the two stimuli to each other than the right half faces.

Third, since there is some evidence in the literature that differences in the performance of men and women occur in some face perception tasks, we balanced the number of male and female participants in our experiments. A female advantage to recognize faces (Lewin & Herlitz, 2002; Rehnmann & Herlitz, 2007; Guillem & Mograss, 2005) and to classify them by sex (O'Toole, Peterson, Deffenbacher, 1996) has been shown, as well as differences in adaptation to the sex of faces depending on observers' own sex (Webster, Kaping, Mizokami, & Duhamel, 2004). We therefore compared the data of these two subgroups of participants, to correlate potential performance differences with different looking behavior in male and female observers.

Since we could control both sides of the comparison, we manipulated the difficulty of the tasks systematically by varying the similarity of the two faces shown in a trial. That way, we could investigate whether changes in difficulty lead to changes in viewing strategy (and if so, if they lead to similar changes in different tasks). Previous findings have shown that, for example, the common difficulty in recognizing other-race faces correlates with a more feature-based processing of those stimuli, compared to more configural processing of own-race faces, i.e., a task that we have naturally more expertise in (e.g., Michel, Caldara, Rossion, 2006; Michel, Rossion, Han, Chung, Caldara, 2006).

The facial manipulations we investigated were related to identity and sex, as these are natural, socially meaningful, and well-studied properties of human faces. Observers performed two same-different tasks with face pairs manipulated in either sex or identity, respectively.

A third task was a 2-alternative-forced-choice task asking participants to point to the more feminine face of each pair. The reason to add this task was that we wanted to know whether facial information is accessed differently when task instruction changes but the stimuli, as well as the facial characteristic of interest, remain the same. For that purpose, we exploited an apparent inconsistency in the face literature about the perception of sex: On one hand, there is the very fast and robust ability to categorize faces by their sex (e.g., Bruce & Young, 1986; Bruyer, Galvez, & Prairial, 1993; Deffenbacher, Hendrickson, O'Toole, Huff, Abdi, 1998); on the other hand, it has been shown that observers are very poor at noticing and remembering variations of sex-related facial information (Bülthoff & Newell, 2004). We hypothesize that the difficulty of the latter task may result from its unnaturalness - people generally do not change their sex, even ever so slightly, on a regular basis. From this, we further assume that these differences in 13

Many of the studies about face perception mentioned so far used small numbers of face images that were derived from normal photographs equated in size and contrast. In our study, control over the face stimuli was improved by using an in-house face database containing over 300 three-dimensional computer-reconstructed laser scans of real male and female heads. This facilitated the generation of a large number of stimuli with homogenous, well-controlled properties (e.g., orientation, size, lighting conditions).

By using the 'flexible face model' of Blanz and Vetter (1999), we could selectively modify specific face properties (such as sex and identity) of our face stimuli. This model represents faces as location vectors in a multidimensional face space (see Valentine, 1991). Thus, manipulation of the position of a face along just one dimension in terms of this face space results in varying only this property while the other characteristics of the face remain unchanged. Another advantage of this flexible model over standard morph-based codes is that it includes correspondences for all sample points (i.e., approximately 70,000 vertices), rather than just a subset of key feature points. This enables us to capture subtle shape and texture variations in the internal regions of the faces and to generate natural looking stimuli.

In summary, the viewing behavior and discrimination performance of male and female observers comparing two faces was assessed. The similarity of the two faces of a pair was varied parametrically within each task, and we assumed that task difficulty would vary accordingly; the faces being compared differed in sex or identity characteristics and in the case of sex being manipulated, two kinds of comparisons were performed. We wanted to test whether the same facial information considered diagnostic in single-face studies was also found when the stimuli to be compared are unfamiliar and remain visible, thus when the information essential for solving the task can be accessed as needed and it is unnecessary to encode or access a detailed internal memory representation. Furthermore, we investigated how observers compared the two faces of a pair, looking for differential scanning patterns depending on the testing conditions. We also explored whether different viewing patterns might emerge when dealing with the same facial characteristic (sex) under different task instructions, by building on a seeming contradiction in the literature about sex perception. Finally, in view of the sex differences reported in face recognition literature we investigated whether male and female participants examine face pairs in a different way.

Methods

Stimuli

Static face images were derived from three-dimensional laser scans collected in the face database of the Max Planck Institute for Biological Cybernetics (http://faces.kyb.tuebingen.mpg.de). The full-frontal two-dimensional face images derived from these heads were between 460 and 550 pixels in height and between 320 and 390 pixels in width, which corresponds roughly to the natural size of a face from chin to hair line (about 20 by 14 cm). At a viewing distance of 56 cm (a distance at which adults typically interact (Baxter, J.C., 1970), a face covered a visual area of about 20 x 14 degrees of visual angle. All faces were shown in color (24-bit color depth).

External features like ears and hair were removed and the faces were free from other cues like make-up or facial hair. To eliminate irregular edges resulting from the removal of the scalp, that could be used as a cue to distinguish the faces, we gradually faded the upper face contour into the background color for each face, by use of a standard alpha blending algorithm with a transparency mask that was empirically determined on the average head.

A morphing algorithm developed by Blanz and Vetter (Blanz &Vetter, 1999; Blanz, 2000) was used to manipulate the faces according to two parameters, i.e., either their identity or their sex, separately. For 'sex morphing', faces were manipulated along the 'sex vector' calculated between the average of all male and the average of all female faces of the scanned population of heads. Nine evenly distributed face morphs (steps of 10% change) were then created between each face identity and its opposite-sex version. Identity morphs were obtained by computing a linear combination of the vectors of two faces of the same sex, creating nine evenly distributed morphs between the two original faces. As each participant performed two tasks with sex-manipulated faces and one task with identity-manipulated faces, more faces (95 female and 95 male) were used to create sex morphs than identity morphs (70 female and 70 male). No face identity was used twice in the whole experiment. *Figure 1* shows two examples of 'morph continua'.

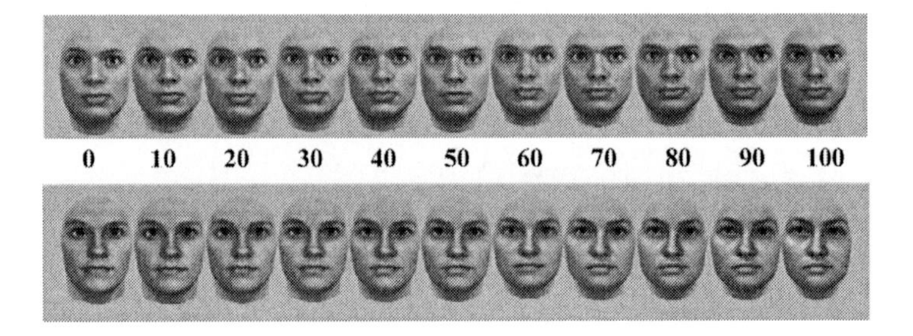

Figure 1: Example of sex (upper row) and identity (lower row) morph continua. The original face ('0') is on the left; numbers indicate the percentage of morphing toward the opposite sex or identity *(see text for more details)*. The face stimuli were presented in color.

Apparatus

The stimuli were presented using E-prime (© Psychology Software Tools, Inc., version 1.1) on a desktop computer linked to a 17-inch CRT monitor (resolution 1024 x 768 pixels, refresh rate 75Hz). Eye movements were recorded with an iView X SMI® dark pupil remote eye-tracker system. Retinal and corneal reflections induced by an infra-red source were recorded at a frequency of 60 Hz to obtain participants' point of fixation on the computer screen (error of resolution: .5° – 1°). Participants were sitting in front of the monitor displaying the test stimuli at a distance of 56 cm and placed their head on a chin rest with a head band to avoid movements. A button box (Psychology Software Tools, Inc.) was used for collecting responses.

Design

In all trials, two face images were presented side by side on the monitor. At the designated distance it was not possible to focus on both faces simultaneously. As previous work demonstrated that visual information retrieving is in general limited to a small region of about 4° around the fixation point (Henderson, Williams, Castelhano, & Falk, 2003), we were sure to prevent participants from perceiving all the information on the screen 'at a glance'. However, the visual angle covered by both faces (about 28° by 20°) was small enough to allow participants to keep their head still when fixating on any part of them. Each face pair consisted of two face images from the same morph continuum (see *Figure* 1 showing a sex continuum consisting of one face in its original appearance and its associated sex morphs). The faces of a pair could be identical or different. The smaller the distance between the 'morph levels' of the two faces in a pair, the more similar the faces were to each other in that trial.

Participants were asked to perform three different experimental tasks: For the 'Sex task' and the 'Identity task' (the same-different tasks) face stimuli from either the sex or the identity morph continua were used, respectively. In the 'Feminine task' (a 2AFC task asking participants to point to the more feminine face of each pair), the same kind of sex morphs as in the Sex task was used, but only different face pairs were shown. Participants performed first the Sex and the Identity task (order counterbalanced across participants). Since the Feminine task was different, it was always performed last: We did not want to affect participants' intuitive strategy in the same-different tasks as a result of their experience with the 2AFC task. To uncover any effect of practice in the Feminine task, we conducted a control experiment (see below).

Each of the tasks was based on a within-subject Repeated Measures (RM) design with 'morph distance' between the faces of a pair as the independent variable; the dependent variables were response time (RT), accuracy (ACC), number of fixations, duration of fixation, time spent within a region of interest, and a range of other eye-position measures that will be specified in the results section.

To rule out face familiarity effects within the tasks, as they might affect not only identification and fixation patterns (see Introduction) but also sex discrimination if both are linked (suggested for example by Rossion, 2002; Bülthoff & Newell, 2004; Ganel & Goshen-Gottstein, 2002), each face morph continuum (original face plus morphs, see *Figure 1*) was used only once in the whole experiment.

In a study using face stimuli derived from the same database and manipulated with the same morphing algorithm, Bülthoff and Newell found that people were not able to distinguish between an original face and a sex morph of that face at a morph distance level of 20% (Bülthoff & Newell, 2004). Therefore, to provide about the same difficulty in the two same-different tasks, and because of time constraints of the eye-tracking technique, we showed only face pairs differing by a morph distance of 30% or greater in the Sex task and the Feminine task.

For both same-different tasks, in half of the trials, face stimuli were paired with themselves ('same pairs'), in the other half the face morphs were paired with the original face of their morph series ('different pairs'). Two stimuli sets were created for the Sex task by using each face once in a 'same pair' trial in one set and once in a 'different pair' trial in the other set. The same procedure was implemented for the faces used in the Identity task. The sets were counterbalanced across participants. In the Feminine task, all faces were paired with the original face of their morph continuum as only 'different pairs' were required. Two stimulus sets for this task were created by assigning each face to one morph level in one set and to another morph level in the second set. For the control experiment (Experiment II), a third stimulus set for the Feminine task was created with faces used before in the Sex and Identity tasks (see below). The Sex (Identity) task consisted of 144 (140) same and different trials. In order to test participants for no more than one hour (longer durations usually result in a drop of eye-tracking accuracy due to fatigue - causing e.g., excessive blinking - of participants), the Feminine task was designed with 44 trials only. Total time of the experiment including instruction, training and calibration procedures (around one hour) was determined in several pilot experiments.

Participants

All participants were between 18 and 37 years old and had normal or corrected-to-normal vision. Most of them were paid volunteers recruited via the MPI Subject Database; a few were members of the Max Planck Institute. All of them were naïve as to the purpose of the experiments and participated only once.

Procedure

Participants were given oral instructions before entering the test cubicle. Instruction consistency was ensured by using a written checklist. A calibration procedure was carried out before the experiment and in between the tasks. After each task, participants left the room for a few minutes to make sure they relaxed. For the same-different tasks, participants were instructed to decide in each trial whether the two faces were identical or not and they were told in which aspect they had to look for differences (identity or sex). In the Feminine task they were told to find the more feminine face of each face pair. No feedback was given during testing but it was provided in a training phase preceding each task (see below). After the experiment participants were asked several questions about their performance (for example 'did you find the tasks equally easy/difficult to solve', 'what do you think is the difference between male and female faces?') to verify that they had performed the tasks correctly. Details of the main and the control experiment are given below.

Experiment I: Thirty-two participants completed this experiment. Fifteen of them were female. Trials were organized as follows: A fixation cross (shown for 750 ms) preceded the test stimulus (two faces side by side). The face images disappeared as soon as a response button was pressed and were followed by a blank screen for 600 ms.

Before each experimental block, participants went through a training block with feedback. Based on the mean response times measured in a pilot experiment (that differed from the main experiment only insofar as there was no time limit) , decision time was restrained to 10 s in the Sex task and to 7 s in the Identity task, while participants had a maximum of 5 s in the Feminine task. These time restraints were designed to avoid extreme outliers in response time but were defined broadly enough not to influence participants' response strategy. If no response was given before the end of that period, the face stimuli disappeared and the trial was excluded from the analysis. Participants were informed about the limited response time and could experience it in the training block. Later in the results section we will present observers' gaze behavior (presented as percentage of fixations to the different AOIs) from Experiment I and from the pilot experiment mentioned above together (*Figure 6*), to demonstrate that there was indeed no effect of time restriction on viewing behavior. Furthermore, the much shorter time to respond in the Feminine task did not lead to a higher loss of trials due to time-outs in this task compared to the other two: Calculated across all participants, 3.9% of trials per task were lost in the Sex task (SEM = 0.7), where observers had the longest time to respond, 3.6% in the Identity task (SEM = 1.0), and 3.7% in the Feminine task (SEM = 1.2) with the strictest time limit.

Experiment II: Seventeen participants took part in this experiment. Eight of them were female. Participants performed only the Feminine task. Only psychophysical data was collected. The response time was limited to 5 s as in Experiment I and participants went through the same training procedure.

Data Acquisition and Reduction

E-prime software recorded psychophysical and stimulus data for every trial. The trials in which participants did not answer within the restricted time frame were not taken into account for the analysis of psychophysical and eye-tracking data. Eye-tracking data were recorded as x and y pixel coordinates of point of regard (POR) on the screen during each trial. Recording started at the beginning of the trial and stopped with the subject's response. Trials with more than 15% missing data points (due to blinks or tracking instability) were excluded from the analysis, as well as participants' data for a whole task if they had more than 15% missing trials. Two participants who reported that their response was based on cues not related to the required execution of the tasks were completely excluded from the analysis. Of the 30 remaining participants, two more were completely excluded for total loss of tracking data; 10 participants' data for all tasks and 18 participants' data for one or two of the tasks could be used. Altogether, we remained with the data of 20 participants for the Sex and the Feminine task (10 feminine, 10 masculine), and 18 participants for the Identity task (8 feminine, 10 masculine). Thus, there was neither a gender skew, nor was the data of one task disproportionately removed from the analysis.

We did not restrain our analyses to only those trials where participants had answered correctly, for two reasons: (1) Trial number (i.e., statistical validity) would have been largely reduced, and, moreover, this reduction would have disproportionately affected the different tasks. (2) We were not interested in the most effective strategy but in the strategy participants considered the most appropriate for a certain task.

Data Analysis and Statistics

Position values for key feature points in the face (e.g., medial and lateral corners of the eyes, tip of the nose, corners of the mouth etc.) were defined on one reference face. Since the faces in our data base are represented in correspondence to each other, we could use the Morphable Model algorithm (see above) to define automatically the same points on each face individually. On the basis of these points, all face images were individually segmented into ten areas of interest (AOIs). See *Figure 2* for a schematic example.

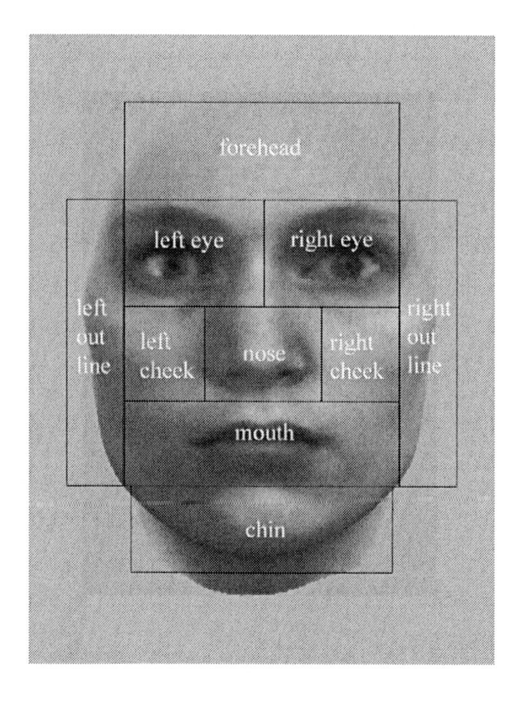

Figure 2: Ten areas of interest (AOIs) were defined on each face individually, using key feature points (e.g., corners of the eyes, tip of the nose, etc.).

We used the freeware MatLab toolbox 'iLab' (Gitelman, 2002) for filtering blinks and calculating fixations from the iView X trial data. A fixation was defined as a set of consecutive gaze coordinates, confined within a diameter of 1° of visual field for a duration of 100 ms or more (Williams et al., 2001). The coordinate values of each fixation determined in which AOI the fixation was made.

All data were analyzed using MatLab (© 1994-2006 The MathWorks, Inc., Version 6.5.1 and 7.1), Excel 2000, and SPSS for Windows (SPSS Inc., Chicago, Illinois, Version 12.0). Once position, time, and length of all fixations had been calculated, a set of analyses was conducted to investigate various aspects of viewing behavior. These analyses will be described along with the results in the next section.

The analyses within tasks were based on the GLM Repeated Measures Analyses (fixed-effects) Model in SPSS. As the tasks differed in paradigms (same-different versus 2-alternative-forced choice) and stimuli used (no 'same pairs' in Feminine task; not all morph distance levels in the Sex and Feminine tasks), their data were primarily compared qualitatively but not quantitatively; i.e., by performing statistical tests within tasks. The more assumption-free non-parametric Wilcoxon Signed Rank test was used to compare across tasks for some explicitly chosen normalized measures.

Results

Psychophysical Data

Repeated Measures (RM) Analyses of Variance were conducted on performance data (ACC) and on response times (RT) for each task separately. The significance level threshold was set to $p = 0.05$. *Morph distance* (i.e., the difference in morph percentage between the two faces shown in a trial, see *Figure 1*) was used as within-subjects factor and *sex of participants* as between-subjects factor. To test if the relationship between the physical manipulation (i.e., a linear operation) and the perception of our stimuli is monotonic (as hypothesized), a linear trend-analysis across consecutive distance levels was calculated (i.e., level 30% – 100% for the Sex and Feminine task, level 10% – 100% for the Identity task). The distance level 0% was not taken into account for the linear trend, because it represents a different kind of stimuli (identical faces) and expected response ("same").

As "correct" trial in the Sex and the Identity task, we counted a "same" response for morph distance level 0, and a "different" response for all other levels (10% - 100%). In the Feminine task, a response was correct if observers indicated the one face of a pair that was closer to the feminine endpoint face (see *Figure 1*, above).

Experiment I

Performance: Results for all tasks are shown in *Figure 3A*. The ANOVAs revealed an effect of *morph distance* for all tasks [all $p < 0.0001$; Sex: $F(8,144) = 30.70$, Identity: $F(10,160) = 56.84$, Feminine: $F(7,126) = 6.47$] but no effect of *sex of participants*. In all tasks a linear trend along consecutive morph distance levels was found [all $p < 0.0001$; Sex: $F(1,18) = 195.48$, Identity: $F(1,16) = 159.84$, Feminine: $F(1,18) = 18.63$].

Response Times: Data are shown in *Figure 3B*. There was an effect of *morph distance* for the three tasks [all $p < 0.0001$; Sex: $F(8,128) = 7.37$, Identity: $F(10,70) = 11.72$, Feminine: $F(7,126) = 4.53$] but no effect of *sex of participants*. In all three tasks a linear trend of RT along consecutive morph distance levels was found [all $p < 0.005$; Sex: $F(1,16) = 44.24$, Identity: $F(1,7) = 8.56$, Feminine: $F(1,18) = 14.00$].

Wilcoxon Signed Rank tests revealed significant differences for both psychophysical measures between all tasks (all $p < 0.0001$; RT: Sex-Feminine: $Z = -10.094$; Identity-Feminine: $Z = -4.535$; Identity-Sex: $Z = -8.982$; ACC: Sex-Feminine: $Z = -8.821$; Identity-Feminine: $Z = -8.088$; Identity-Sex: $Z = -5.0640$).

To summarize, in all three tasks male and female observers were better and faster when the two faces on the screen were at a large morph distance and their performance became worse as well as slower with decreasing morph distance. Thus the similarity of the faces of a pair was indeed monotonically related to the linear manipulation in physical space, allowing us to directly relate the fixation measures to this factor as well (see below). Performance was best in the Feminine task; observers were less good in the Identity task and worst as well as much slower in the Sex task. It seems worth mentioning again that the face stimuli were manipulated in exactly the same way in the Sex and the Feminine task.

Experiment II

In Experiment I, all participants performed the Feminine task last. Experiment II was designed to test whether task order might influence performance and thus explain superior performance in the Feminine task. All participants performed only the Feminine task in Experiment II.

The mean percentage of correct responses to all stimuli pairs across participants is shown in *Figure 4* as 'Fem^{1st}'. The performance data of the Sex task in Experiment I was split into two groups: 'Sex^{1st}' and 'Sex^{2nd}', which represent data of those participants who had performed the Sex task first and second (i.e., after the Identity task), respectively. Similarly, 'Fem^{2nd}', in *Figure 4* shows the performance for the Feminine task in Experiment I. Even though there is some within-experiment training effect, as participant's performance for the Feminine task is worse in Experiment II than in Experiment I, it clearly surpasses performance for the Sex task (Wilcoxon: Fem^{1st} – Sex (both groups): $p = 0.000$, $Z = -8.839$). In consequence, as the role of training has been refuted, the question remains whether this striking difference in performance between the Sex and the Feminine task is paralleled and might be elucidated by distinctive eye movement patterns, analogous, for example, to what has been shown for differing facial information processing in same-race and other-race faces (see Introduction).

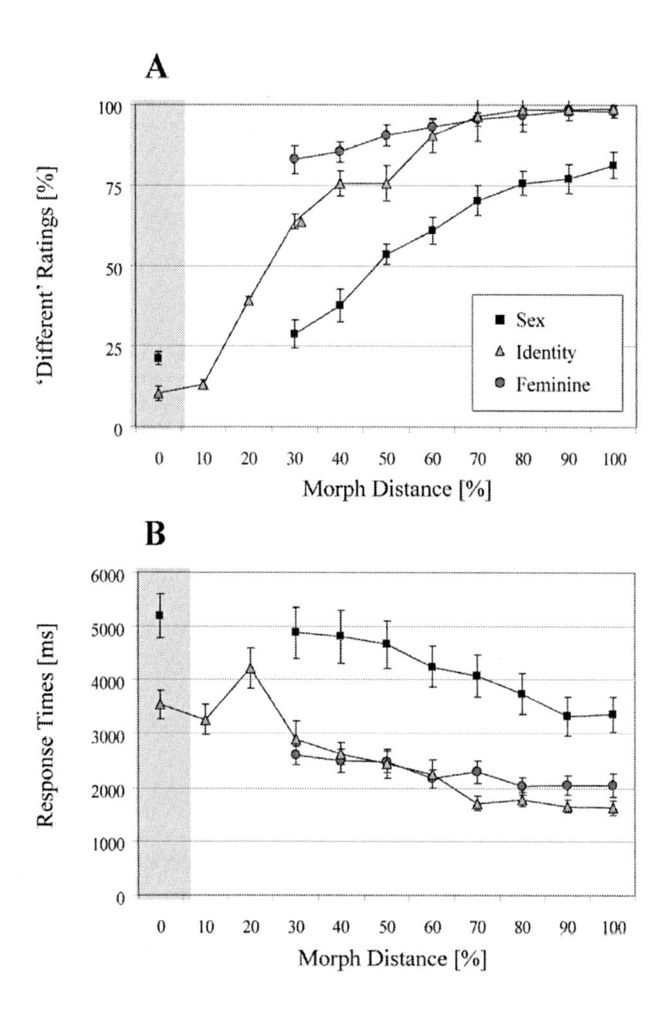

Figure 3: Psychophysical data for Experiment I. A: Performance, shown as mean percentage of 'different' ratings, equivalent to correct responses for all morph distance levels except 0% ('same pairs', dashed columns). Note that in the Feminine task, no such 'same pairs' were presented, and that in the Sex and Feminine task no face pairs at the morph distance 10% or 20% were used. There were as many same trials as different trials in the Sex and the Identity task. N = 20 (Sex and Feminine task), and 18 (Identity task). Error bars represent SEM. B: Response Times in ms.

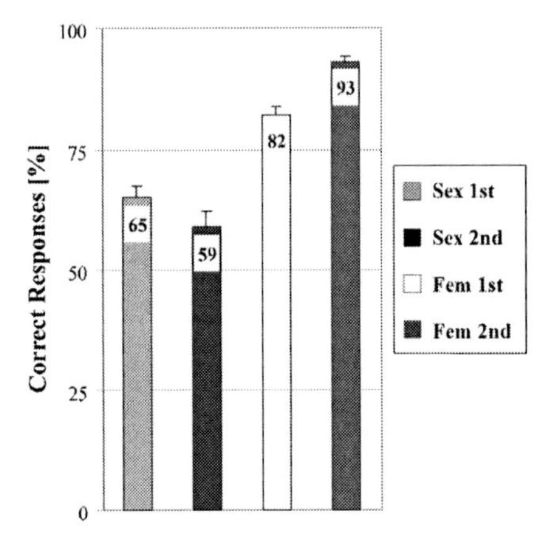

Figure 4: Control for training effects. Mean number of correct responses for participants who did the Sex task first in Experiment I (Sex^{1st}, N = 12, 6 male, 6 female), or after the Identity task (Sex^{2nd}, N = 8, 5 male, 3 female). Fem^{1st} shows accuracy for participants who did only the Feminine task (Experiment II, N = 17, 8 male, 9 female), Fem^{2nd} for those who did the Feminine task at the end of Experiment I (N = 20; 10 male, 10 female). Error bars represent SEM.

Eye-Tracking Data Analysis

RM Analyses of Variance were conducted for each task separately. The significance level threshold for all statistical analyses was set to p = 0.05. For some global measures, tasks were compared with the non-parametric Wilcoxon Signed Rank test to account for different task designs.

Single fixation duration in direct comparison. Longer fixations imply that the user is spending more time interpreting or relating the observed visual stimulus to internalized representations (Goldberg & Kotval, 1999). Buswell (1935) found that observers made longer fixations with increasing task difficulty. RM Analyses of Variance with *morph distance* as within-subjects factor and *sex of participants* as between-subjects factor revealed no significant effects, as fixation duration did not vary with difficulty. This is in accordance with our assumption that in a comparison task, there is no need to internally relate stimulus and representation since all the information necessary to solve the task is visually available and can be inspected as often as necessary. The mean length of a single fixation was not significantly different in the three tasks, ranging between 351 ms (Identity task, SEM = 22) and 382 ms (Feminine task, SEM = 24; Sex task = 367 ms, SEM = 23). Since single fixation duration did not vary across AOIs either, the distributions of numbers of fixations and of cumulative fixation time in the different AOIs are highly correlated and thus redundant. We will therefore only present results related to the number of fixations.

Overall number of fixations per task. *Figure 5* shows the overall fixation number per trial for each task. Overall, observers made most fixations (and thus spent most time fixating) in the Sex task, followed by the Identity task and then the Feminine task (Wilcoxon Signed Rank: all $p < 0.005$; Sex-Feminine: $Z = -8.452$; Identity-Feminine: $Z = -3.033$; Identity-Sex: $Z = -5.862$). These findings correlate with the response times for the tasks (see above), showing that the worse and the slower the performance, the more fixations are made (and the more time is spent looking at the faces).

That the relationship between accuracy, reaction time, and fixation time was not caused by different levels of observers' attention is indicated by similar values of ***search efficiency*** in the three tasks. It is calculated as the number of fixations on both faces divided by the number of all fixations made on the screen during a trial (Goldberg & Kotval, 1999). A value of 1 would mean that no fixation was made outside the face stimuli (which covered about 28° x 20° of a 40° x 30° screen), while smaller ratios indicate lower efficiency. Search efficiency was highest for the Sex task (0.92, SEM = 0.07), and slightly lower for the Identity (0.89, SEM = 0.06) and Feminine task (0.86, SEM = 0.04), thus following a trend inverse of that of accuracy.

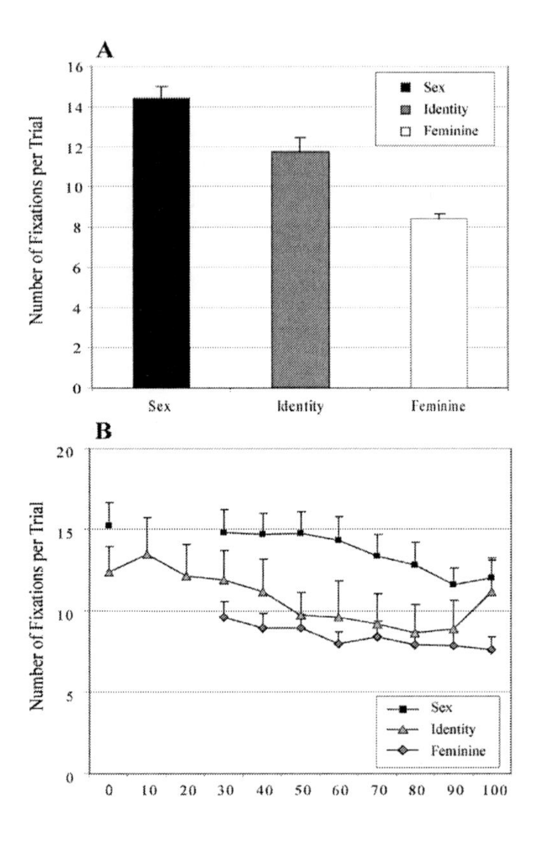

Figure 5: Mean number of fixations for each task on average (A) and across morph levels (B). Error bars represent SEM. N = 20 (Sex and Feminine task), and 18 (Identity task).

<u>General analysis of eye movement patterns</u>

RM Analyses of Variance with within-subjects factor *face* (2 levels, right or left on the screen), factor *AOI* (10 levels per face), and factor *morph distance* (9 levels for Sex, 11 levels for Identity, 8 for Feminine) were conducted on numbers of fixations. Between-subjects factor was *sex of participants*. See *Table 1* for complete F - and p - values. In the following, the results will be described along the three main questions that have been raised in the introduction: (1) Which facial features are considered essential to solve the tasks? (2) How do observers compare the two faces? (3) Do the results differ for male and female participants?

(1) Which features do play a role? Does the Task make a difference? The numerous areas of interest on the faces received unequal numbers of fixations, resulting in a significant effect of AOI on fixations for all tasks. *Figure 6* shows the overall fixation pattern in AOIs (collapsed across features and both faces) for the pilot study (*Figure 6A*) and for Experiment I (*Figure 6B*), demonstrating that there was no effect of the differing time limits on looking behavior: Even though only few participants performed this pilot experiment, the eye movements to the areas of interest already show a general pattern that corresponds to the one in the main experiment (note especially the eye and nose regions). See figure legend for numbers of participants. In the following, we will only refer to the data of Experiment I (*Figure 6B*).

Participants fixated most often and longest on the eyes and the nose in all tasks, followed by the mouth in the Sex task and the cheeks in the Identity and Feminine task. The eyes were more often looked at in the Sex task (on average, 36.8 % of all fixations in a trial) than in the Identity (31.6 %) and the Feminine task (28.8 % of all fixations). The eyes received significantly more fixations than all other AOIs except the nose in the Sex task only ($p < 0.003$, Bonferroni-corrected post-hoc pair-wise comparisons). The nose was most looked at in the Feminine task (32.0 % of all fixations, versus 26.3 % in the Sex and 24.5 % in the Identity task); unlike in the other tasks, in the Feminine task the nose was even slightly more looked at than the eyes (although the differences between eyes and nose were never significant). Wilcoxon Signed Ranks tests between tasks revealed no significant difference for fixations to the eyes or the nose region.

Observers spent more fixations on the mouth area in both same-different tasks than in the Feminine task (only 7.5 %, versus 13.44 % and 14.45 %; Wilcoxon: Feminine – Sex: p = 0.041, Z = -1.904; Feminine – Identity: p = 0.020, Z = -2.330). The cheeks were looked at less often in the Sex task (11.24 %) than in the other two tasks (Identity: 17.99 %, Feminine: 19.44 % of all fixations; Wilcoxon: n.s.). Numbers of fixations to the face outline were highest in the Feminine task and lowest in the Sex task (Wilcoxon: Feminine – Sex: p = 0.000, Z = -3.883; Identity – Sex: p = 0.028, Z = -2.199). In sum, our first eye movement-related questions can be answered positively: Some facial features are effectively considered more diagnostic than others and which feature is looked at more often depends on the task.

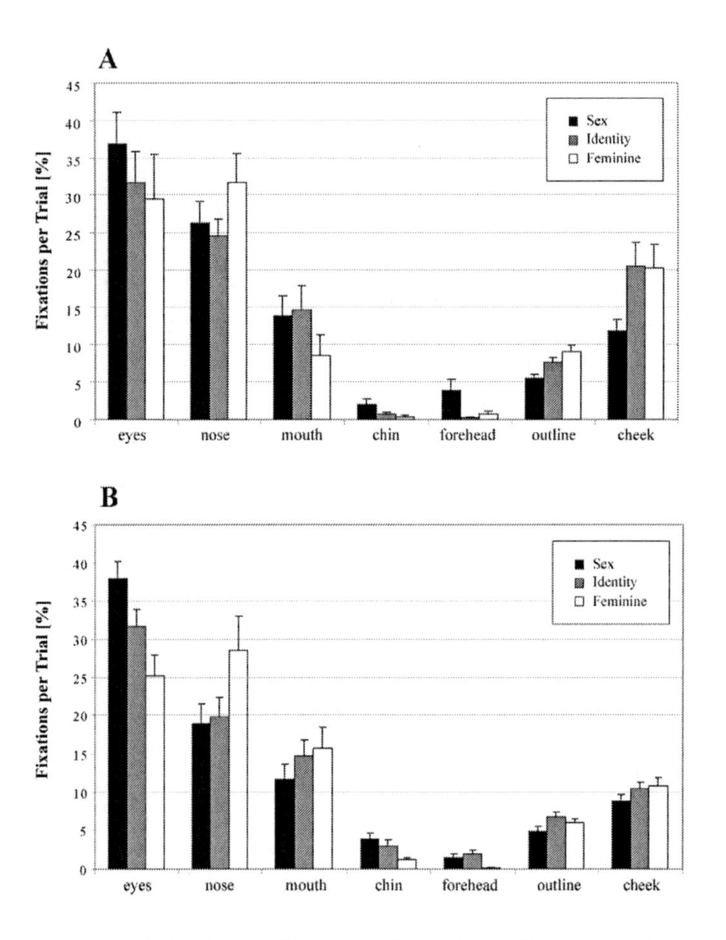

Figure 6: Number of fixations in Areas of Interest. 6A: Pilot experiment without time restriction (N = 11 Sex task, 10 Identity task, 8 Feminine task). 6B: Experiment I (N = 20 Sex and Feminine task, 18 Identity task). Mean percentages of fixations per trial are shown for each task. Fixations are collapsed over features that were represented by more than one area of interest; e.g., 'eyes' represents both eyes of both faces on the screen. Error bars represent SEM.

Where to look first? We analyzed the data of each task with respect to first fixations and subsequent fixations; the results for first fixations are shown in *Figure 7A* as percent of first fixations in each trial averaged across participants. As the results for subsequent fixations practically do not differ from results for all fixations, we do not present additional figures for subsequent fixations. When we collapse the data for first fixations across corresponding AOIs and both faces (not shown), a pattern appears that is similar to that of the data for subsequent and/or all fixations (see *Figure 6* for results calculated over all fixations). For a closer look at the data, in *Figure 7*, we now present numbers of fixations for each single AOI for first fixations (*7A*) and all fixations (*7B*), respectively. The most striking finding here is that in the Feminine task, observers first fixation fell much more often on the nose region of the left face than in the other two tasks (Wilcoxon Signed Ranks: Feminine – Sex: $p = 0.001$, $Z = -3.454$; Feminine – Identity: $p = 0.000$, $Z = -3.610$), whereas there was no difference between the Sex and the Identity task. Even though there are other minor variances between tasks concerning the first fixation in a trial, this most substantial difference for the nose region is in line with the results that we have reported so far about more fixations on average on this region in the Feminine task compared to the other two tasks.

What if the task gets more difficult? The effect of *morph distance* was significant for all tasks, as well as the interaction *AOI*morph distance*. There was a linear trend of numbers of fixations along consecutive *morph distance* levels in all tasks, due to continuously decreasing numbers of fixations with decreasing difficulty (see *Figure 5B*). This is again in accordance with the relation we found between task performance and fixation time (see *Figure 3*). Thus participants in general made more fixations the more difficult a trial, or the whole task was.

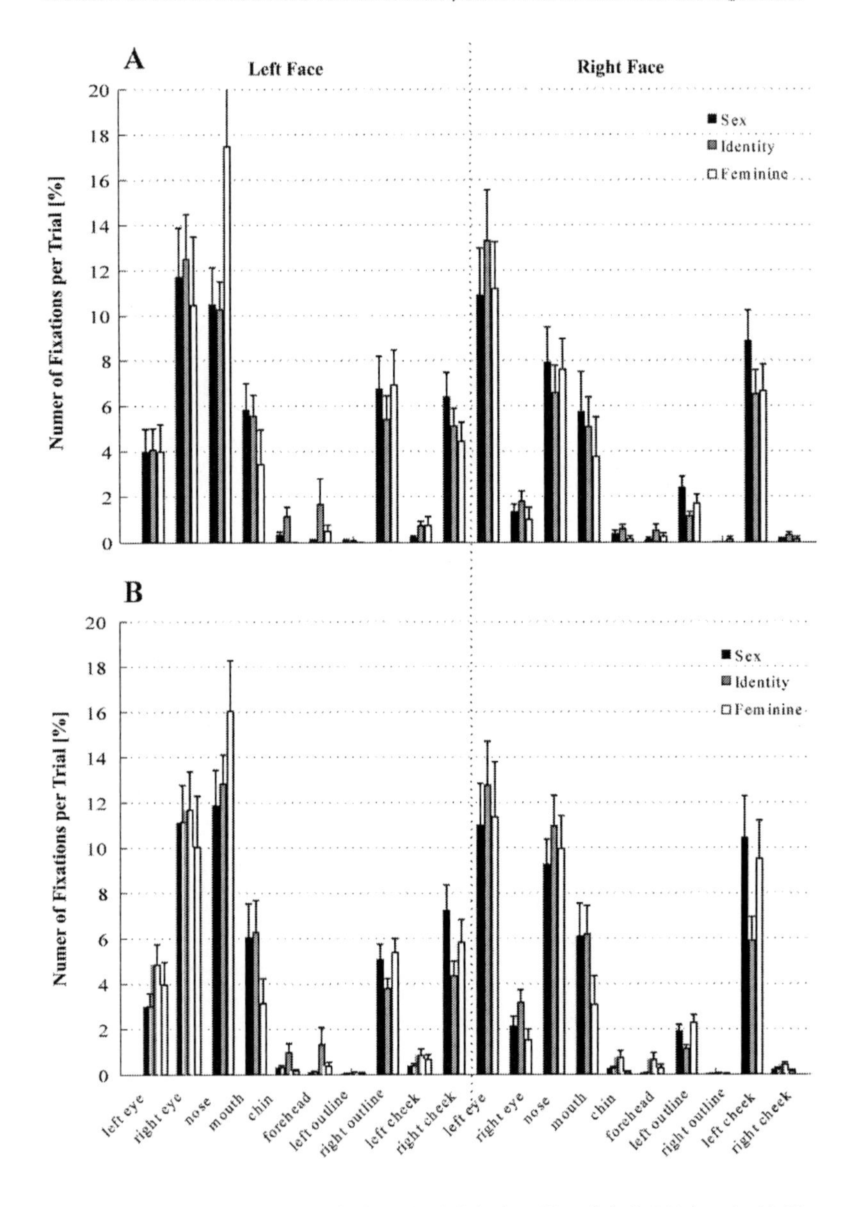

Figure 7: Mean percentages of the first fixation (A) and all fixations (B) to all single AOIs in each trial. (N = 20 Sex and Feminine task, N =18 Identity task). Error bars represent SEM.

When looking closer at the linear relationship between numbers of fixations and difficulty, we found that, with decreasing *morph distance* between the two faces in a trial (i.e., increasing similarity), more fixations were made only on a few AOIs. In each AOI, we calculated one-way RM Analyses of Variance including linear trend-analyses for the numbers of fixations along consecutive morph distance levels. AOIs showing a significant effect for the ANOVA and the linear trend (and thus contributing to the interaction AOIs*morph distance) are shown in *Figure 8*. The 'inner' eye regions (i.e., the eyes closer to the middle of the screen), the nose, and the mouth of the left face show this morph distance effect in the Sex task (all $p < 0.04$ for the linear trend). In the Identity task, the effect is found as well in the inner eye regions, the nose and both mouth regions, and additionally in the left outline and the left cheek of the right face (all $p < 0.01$). In the Feminine task only the numbers of fixations on the nose and mouth area significantly increase with decreasing *morph distance* (all $p < 0.012$).

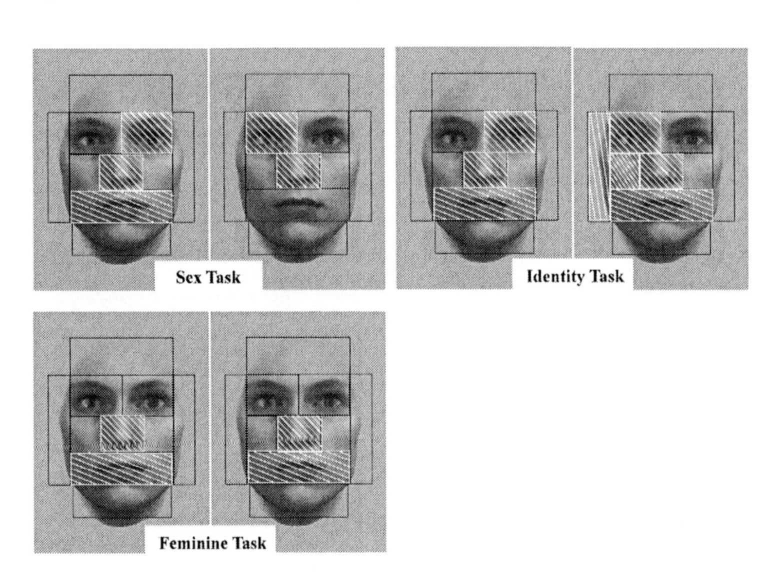

Figure 8: Interaction Areas of Interest*morph distance. Only in few AOIs the number of fixations increased significantly with increasing similarity between the faces in a trial. RM Analyses of Variance and linear trend analyses along morph distance levels were conducted within each AOI separately. The AOIs in which effects of both analyses were significant are highlighted for each task.

It appears from these results that when task difficulty increased due to increasing face similarity (i.e., decreasing morph distance) within a task, observers made more fixations only to a few AOIs, and did not distribute them evenly over the faces. Those AOIs receiving increased scrutiny differed slightly depending on the task. The most obvious difference is that the eyes, while being the most looked at facial features in the whole experiment, are not considered informative in the Feminine task when observers are facing higher task difficulty.

(2) How are two faces compared? The ANOVA revealed a significant effect of *face* (left or right of a pair) only for the Feminine task; this is due to a higher number (18.9%, SEM = 1.9) of fixations per trial on the left face compared to the right face (12.72%, SEM = 1.7) only in this task. Consistently, of the first fixations in the Feminine task (see *Figure 7*, above) 41.6% (SEM = 1.6) fell on the left face but only 32.5% (SEM = 1.2) on the right face (this discrepancy is mainly due to the very high percentage of first fixation to the left nose), while this difference was much smaller (Sex task: 37.9% left, SEM = 0.9; 35.9 right, SEM = 1.4) or nonexistent (Identity task: 37.9% left, SEM = 1.1; 37.8% right, SEM = 1.3) in the other two tasks.

The **ratio of transitions** between the two faces was calculated for each task by counting the gaze transitions from one face to the other in each trial and dividing this number by the sum of all fixations made in the trial. The maximum ratio of 1 indicates that gaze was moved from one face to the other after each single fixation. This measure was assessed to give insight into how participants compared the two faces in general. No significant differences or meaningful trends were found within tasks for this measure. On average, participants made approximately two fixations on one face before moving their gaze to the other face in all tasks (mean ratio of transition was 0.44 in the Sex task (SEM = 0.00) and in the Feminine task (SEM = 0.01), and 0.47 in the Identity task (SEM = 0.01)).

Is the left face half more informative? The interaction AOI*face was significant for all tasks, suggesting that observers directed their attention to different areas of interest on the two faces in a trial. *Figure 9* reveals differences between numbers of fixations on AOIs that exist on both sides of each face, for example the eyes. For further analyses, we therefore separated the AOIs into two categories: Mouth, chin and forehead belong to the 'central' AOIs, as they are present only once in a face. Eyes, cheeks, and facial outline exist as left and right AOIs on each face and thus represent the 'bilateral features' (compare *Figure 2* to *Figure 9*).

The Bonferroni-corrected post-hoc pair-wise comparisons that were conducted between all 20 AOIs subsequent to the general RM ANOVA revealed that the numbers of fixations spent on the 'central features' did not differ between left and right face in a trial, except for the nose in the Feminine task ($p < 0.05$), where the left nose was more often and longer looked at than the nose of the right face (6.18 % of fixations more on left nose). Thus apparently fixations on the nose were responsible for the main effect of the factor *face* in the feminine task (see above).

In contrast, we found significant differences for numbers of fixations on most 'bilateral features'. In the Sex and Identity tasks, within each face, fixations on both eye areas differed, while between faces, fixations on the left eyes of both faces differed from each other, and the same effect was found with fixations on the right eyes of both faces (all $p < 0.02$). In the Feminine task, where the eyes received fewer fixations in relation to the other features (see above), the same differences in numbers of fixations were observed but did not reach significance (see *Figure 9*). For the other bilateral features (left and right cheeks, left and right outline), the same pattern was found as for the eyes, and comparisons between and within faces were significant in all tasks (all $p < 0.032$).

To conclude from these results and the graphs shown in *Figure 9*, in all tasks, the bilateral features on the *inner* half of the faces (i.e., closer to the screen center) were looked at more often than the features of the outer half of the faces.

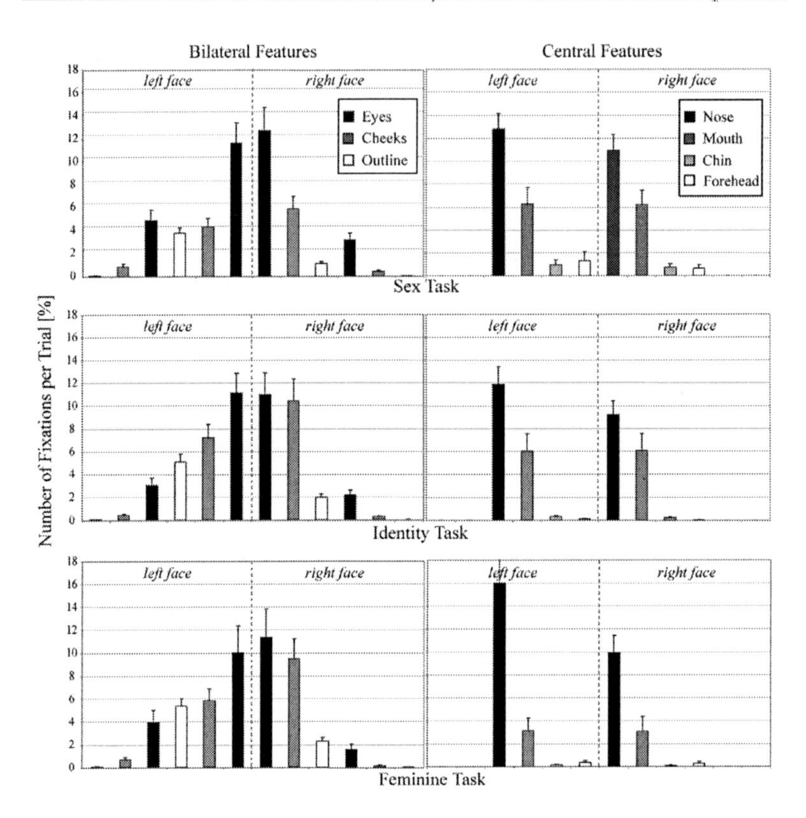

Figure 9: Asymmetries in viewing pattern. AOIs grouped as 'bilateral' features (appearing twice on a face, left) and 'central' features (only one per face, right). Numbers of fixations on central features do not differ between left and right face, except for the nose in the Feminine task. Numbers of fixations on bilateral features differ significantly between faces and between both regions on one face. Error bars represent SEM.

Scan-path strategies: Is there a stereotyped pattern? Viewing behavior can be described as a temporal sequence of successive fixations in different AOIs. We investigated whether the observed sequences show similarities across morph distance levels in each task. To examine the global predictability of scanning, that is, how much the whole scanning sequence of the faces of one trial resembled that of another trial, we did a string-edit comparison of all trial sequences within each task. To this effect, we computed the 'Levensthein distance', a measure borrowed from DNA sequence analysis to calculate the 'editing cost' (how many elements need to be changed) of transforming one sequence into another (Michael Gilleland, http://www.merriampark.com/ld.htm).

We found that the average Levensthein distance between all sequences divided by the average length (i.e., the number of fixations) of the two sequences being compared varied only slightly yet in accordance with the difficulty of the tasks, as follows: It is highest for the Sex task (0.41), slightly lower for the Identity task (0.40) and lowest for the Feminine task (0.36). Thus viewing patterns were slightly more diverse in the more difficult tasks. The results are more meaningful for the variation of sequences within each task: Linear trend analyses across consecutive morph distance levels show that the Levenshtein distance significantly increased with difficulty (i.e., decreasing morph distance) within all tasks ($F > 21.00$, $p < 0.001$ for all tasks). In addition to the similarity between scan-paths, we then searched for distinct patterns in the fixation sequences, as follows.

Do we look again? First, immediately **repeated fixations** on the same region during a trial were analyzed, as indicator of local feature-based processing, opposed to the generation of a global face percept (Barton et al., 2006). The results do not contribute any new information (neither significant effects nor meaningful trends in any of the factors we investigated); it is only worth noting that repeated fixations happened very rarely throughout the whole experiment, i.e., in general after fixating on one AOI, observers' eye gaze moved to another AOI.

Which eyes are compared? More interesting results were obtained when we scanned the eye movement sequences more precisely for **successive eye fixations** (i.e., successive fixations from one eye region to another eye region on the same or other face), as eyes are the bilateral facial features that were looked at most frequently in general throughout the experiment. We calculated first the ratio of all successive eye fixations (see *Figure 10*), that is, the number of fixations on one eye region followed by one to any other eye region, in relation to half the number of all eye fixations in a trial (i.e., all possible repetitive fixations). No difference between tasks was found for this measure.

Then, in the same way, the ratio of 'symmetrical' successive fixations was calculated, that is, from left/right eye of the left face to the left/right eye of the right face and vice versa, as well as the ratio of 'asymmetrical' successive fixations, e.g., from right (inner) eye on the left face to the left (inner) eye on the right face. As is clearly visible in *Figure 10*, observers compared the eyes of a face pair asymmetrically most of the time, moving their gaze from inner eye to inner eye or (less often) from outer eye to outer eye. This clarifies the fixation pattern shown in *Figure 9*, where the features/eyes closer to the center of the screen were looked at more often than the outer features/eyes of both faces.

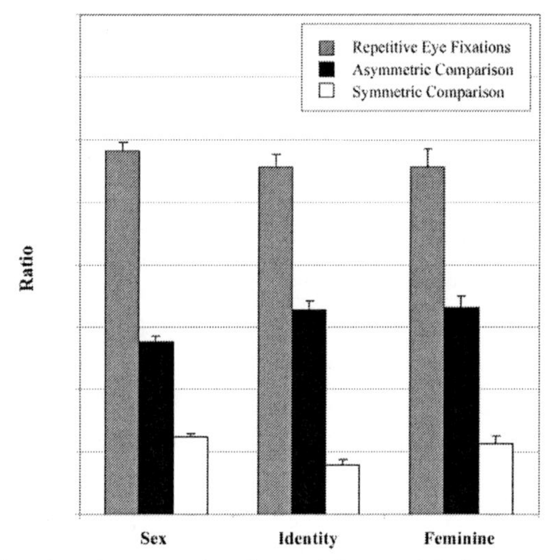

Figure 10: Grey bars show the mean number of successively repeated fixations to any eye area per trial divided by half the number of fixations made in that trial, i.e., the number of possible repetitive fixations. The same ratio was calculated for 'asymmetrical comparisons' between the eye regions of both faces (black bars), and 'symmetric comparisons' (white bars). Error bars represent SEM.

(3) Does observer's sex make a difference? No main effect of *sex of participants* was found in any of the tasks, but an interaction *AOI*sex* of participants was present in the Sex task and the Feminine task. *Figure 11* shows the mean percentage of fixations on AOIs (collapsed across bilateral features and both faces) for male and female participants separately for these two tasks. The difference between male and female observers' results are the following: Female participants made on average more fixations to the eyes of the faces than male participants. Independent-samples t-tests that were conducted to compare fixating behavior of male and female observers to each AOI revealed significant differences for the eye regions in both tasks ($p < 0.025$). Additionally, male participants made on average more fixations to the cheeks than female participants, although this difference between participants reached significance only in the Feminine task ($p = 0.048$).

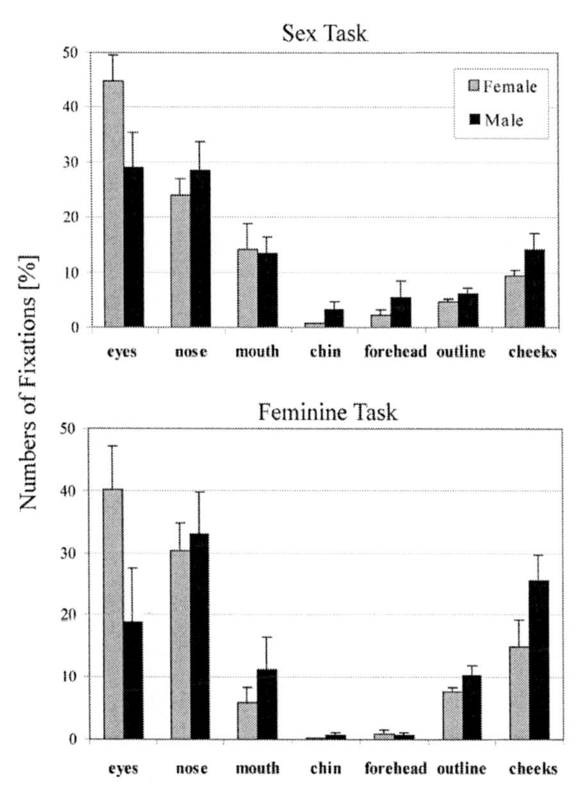

Figure 11: Mean percentages of fixations per trial per Area of Interest, shown for male and female participants separately. Features are collapsed across bilateral features and both faces. N = 20 in both tasks.

Table 1

RM ANOVA on fixation numbers with factors face, AOI, distance level (within) and sex of subject (between)

Factor	Task		
	Sex	**Identity**	**Feminine**
Face (left / right)	not significant	not significant	F(1,18) = 10.123 p = 0.005
AOI (10 areas per face)	F(9,162) = 21.329 p = 0.000	F(9,144) = 10.155 p = 0.000	F(9,162) = 14.125 p = 0.000
Morph Distance (similarity of faces)	F(8,144) = 12.383 p = 0.000	F(10,160) = 9.914 p = 0.000	F(7,126) = 7.207 p = 0.000
Face*AOI	F(9,162) = 21.107 p = 0.000	F(9,144) = 18.590 p = 0.000	F(9,162) = 19.231 p = 0.000
AOI*Morph Distance	F(72,1296) = 2.532 p = 0.011	F(90,1440) = 2.542 p = 0.030	F(63,1134) = 3.282 p = 0.001
AOI*Sex subject	F(9,162) = 2.897 p = 0.034	not significant	F(9,162) = 2.616 p = 0.008
Linear Trend along Morph Distance	F(1,19) = 37.594 p = 0.000	F(1,17) = 34.620 p = 0.000	F(1,19) = 20.578 p = 0.000

Discussion

Performance

The two main findings for the performance data are, first, that unsurprisingly participants were better and faster the more different the faces in the test pair were, and second, that there were large differences in performance between the three tasks. We will review these two findings before considering the data on eye movements.

In all tasks, test pairs of linearly increasing similarity increased task difficulty in a monotonically related way. While the fact that more similar stimuli are more difficult to discriminate is unsurprising, it is worth noting that no sudden changes in performance were observed as a linear relationship between performance (or response times) and morph distance was established. These findings are in agreement with previous studies (Bülthoff & Newell, 2004; Angeli, Davidoff, & Valentine, 2001) on categorical perception that failed to report a sudden change in performance, a hallmark of categorical effect, in unfamiliar typical faces when sex or identity of the faces was manipulated.

Performance varied considerably between tasks; the Sex task was clearly more difficult than the Identity task, whilst the Feminine task was the easiest to solve. In the Identity and Sex tasks, participants performed the same type of judgment (same or different) with face stimuli manipulated along two different dimensions. The results show that - for corresponding morph distance - performance was always better and faster for face pairs in the Identity task. These findings are in accordance with a previous study by Bülthoff and Newell who reported discrimination between faces differing in sex to be more difficult than for faces differing in identity (Bülthoff & Newell, 2004). One might assume that performance in the Identity task is better simply because identity morphs differ perceptually more from each other than sex morphs. However, this assumption is not supported by the other results of the experiment, as performance – in comparison to the results of the Identity task - was better for the Feminine task and much worse for the Sex task although in both of these tasks faces underwent identical sex manipulations. In low-level psychophysics it has been reported repeatedly that performance in a 2AFC-task is in general somewhat higher than in a same-different task (e.g., Macmillan, Goldberg, & Braida, 1988; Chen & Macmillan, 1990; Creelman & Macmillan, 1979; all in: Macmillan & Creelman, 1991). In our study, for face pairs at identical morph distance levels, for example at a 30 % distance, participants were unable to tell that the faces differed (i.e., there were less than 30% correct responses for this level) in the Sex task, but when they knew that the faces differed (in the 2AFC-Feminine task), they could correctly judge which one was more feminine in more than 80 % of the cases. In our opinion, this difference in performance is too high to be a mere consequence of the "task factor" mentioned for low-level psychophysics. We think that another factor influencing observers' discrimination capacity might be differences in attention allocation caused by the instructions for the task at hand. Participants apparently were induced to look at the stimuli in a way that was adequate to allow a correct answer in the Feminine task but not in the Sex task. Schyns and colleagues have suggested that observers direct their attention to different spatial scales depending on the facial information they are looking for (Schyns & Oliva, 1999; Goffaux et al., 2005). Our findings that observers' fixation behavior towards the face stimuli was indeed different in the two tasks (this will be discussed in the following section on eye-movements) might be a reason for participants' huge performance differences, in agreement with Schyns' suggestion.

Since the Feminine task was always performed last by all participants, the question remained whether the surprisingly better performance observed for the Feminine task might result from implicit learning due to repeated exposure to the sex-morphed stimuli. The results of Experiment II (where observers performed only the Feminine task) show that this is not the case. Although observers' performance was reduced compared to Experiment I, it was significantly higher than performance in the Sex task of Experiment I, irrespective of whether participants had performed the Sex task first or second in that experiment (see *Figure 4*). The effect cannot be due to particular face stimuli being accidentally easier to discriminate than others, since Experiment II was done with face stimuli that have been used in the Sex or Identity task of Experiment I before.

Gaze Behavior

Internal representations. In accordance with our suggestion that in a comparison task, there is no need to internally relate the stimulus to any stored representation, we did not find variations in single fixation length related to task or difficulty. This corroborates the argument stated in the introduction, i.e., that the facial features found essential to solve our comparison tasks pertain to a more general concept of identity and sex, rather than to a memory recall of specific faces necessary to solve the task at hand.

Diagnostic Features. Our results confirm earlier findings reporting that not all facial features are equally important for face processing (e.g., Schyns et al., 2002). We have shown that the number of fixations made to different areas of interest varies strongly. This result cannot be related to the size of the areas, as, for example, the largest region is the forehead (covering about 28 % of the whole face), which received very few fixations on average, whereas the eyes were looked at most frequently, even though each of them represents only about 9 % of a face surface. The most 'important' regions regarding the fixations made to them were the eyes and the nose, in all tasks, followed by the mouth (Sex task) and the cheeks (Identity and Feminine task). In contrast to previous studies (e.g., Pearson et al., 2003), the forehead was not extensively scanned in our experiment. Presumably, as our face stimuli fade into the background and no hair line is visible, the forehead area does not provide much information about identity or sex.

Our results emphasize the task-dependency of observers' intention (whether conscious or unconscious) to consider some facial features as more informative and worth of attention than others. In both same-different tasks, the eyes were considered the most informative features (*Figure 6*). Nevertheless, the importance of the eyes in the Sex task was even more highlighted compared to all other features (i.e., they were more often looked at) than in the Identity task, where, as we have shown, observers distributed more fixations over the whole faces. This finding about differing diagnostic features in the Sex and the Identity task is in accordance with recognition and sex classification studies mentioned earlier (Pearson et al., 2003; Schyns et al., 2002) and is also reflected in our analysis on increased numbers of fixations to only some specific areas of interest when a task gets more difficult (*Figure 8*): Here again, in the Identity task, more fixations are distributed to a higher number of regions besides eyes and nose than in the Sex task. Our data corroborate what has been reported so far about diagnostic features in single-face studies involving face identification and sex classification of familiar faces. Since our study was specifically designed to eliminate the role of any task-related memory representations, we can now conclude from our results that these features are truly related to the 'general concept' of identity and of the sex of faces.

In the Feminine task, observers' viewing behavior differed slightly but consistently from their behavior in the two same-different tasks: Not only the nose received on average slightly more fixations than the eye regions; but also the number of fixations significantly increased for the nose and mouth areas but not for the eyes when the difficulty of the task was increased, thus the eyes were not considered as informative as in the other tasks. Additionally, the left face, or, more precisely, the nose of the left face, was looked at significantly more often than the right face (or the nose on the right face); the first fixation in each trial was also by far more often directed to the (left) nose in the Feminine task compared to the other two tasks (see *Figure 7A*). These findings suggest that although observers referred to the sex of the face stimuli in both sex-related tasks (Sex and the Feminine tasks), they distributed their attention (i.e., their eye movements) differently over the face stimuli in both tasks. Hence the question arises whether these differences in viewing behavior can account for the huge discrepancy in performance between the Sex and the Feminine task – this issue will be addressed in the following paragraphs.

Different viewing strategies? The strategy that participants pursued in the Sex task of the present experiment seems to be sub-optimal, if one considers the performance for all tasks, especially the high accuracy in the Feminine task. It would be interesting (but it is, however, beyond the scope of this study) to test directly if participants can be instructed to use a specific strategy, and how that would affect their behavior. We will now discuss in more detail what the differences in strategy might relate to, in view of the behavioral differences we found in our analyses, and in view of other findings from related studies.

Our data show that the nose is the most scanned feature in the Feminine task.

The nose could indeed be diagnostic for the characterization of the sex of faces as proposed by some authors (Roberts & Bruce, 1988; Chronicle, Chan, & Hawkings, 1995; but for different opinions see e.g., Schyns et al., 2002; Pearson et al., 2003). Yet, in our view, it is unlikely that observers used different features in two sex-related tasks performed during the same experiment. Moreover, as Chronicle and colleagues report in their 1995 paper, in full-frontal views the distinctive shape of a nose is not completely available to the perceiver, making sex judgments in that orientation harder rather than easier. We would thus not expect that looking particularly at the nose should lead to notably higher performance in a sex-related task asking for the more feminine of two frontal-view images of faces.

As we already mentioned, different gaze behavior could indicate that observers directed their attention to different spatial scales (Schyns & Oliva, 1999; Goffaux et al., 2005). Schyns and Oliva showed in their 1999 paper flexible use of scale information for different face tasks; amongst other findings they reported that both coarse and fine scales allow for 'gender decisions'. Since the nose coincides with the center of the face, we propose that participants show a stronger preference for this region than for the eyes in the Feminine task because they compared the faces in a more holistic (or 'configural') fashion (i.e., using low spatial frequency information in the visual periphery) rather than by investigating single features (at high spatial frequency); they hence fixated the center of the stimuli more often to get a 'global' view of the faces. The assumption of a holistic approach is substantiated by the fact that, in the post-experimental interviews, most participants reported that they had performed the Feminine task in a more 'intuitive' way than the other two tasks and tried to get an 'overall impression' of one face before comparing it to the other. This orally reported strategy is further corroborated by the considerably higher number of first fixations in each trial to the nose region in the Feminine task (and to the region between the two faces) and could reflect the processing of more low-spatial frequency information compared to more high-level local information in the other tasks, an interpretation that is reminiscent of the findings of Schyns and Oliva as well as Goffaux and colleagues (Schyns & Oliva 1999; Goffaux et al., 2005).

Results of previous studies remain mute about the question of how eye movements differ between holistic and feature-based perception (e.g., Farah, Wilson, Drain, & Tanaka, 1998; Tanaka & Sengco, 1993). We suggest that the viewing behavior observed in the Feminine task, i.e., participants fixate in the middle of one face, presumably to get an overall configural 'impression' at low resolution, and then switch to the other face to do the same, reflects a holistic viewing strategy. We are not the first to assume that the difference between a same-different and a 2AFC task using the same stimuli is a more holistic processing of the face stimuli in the latter: Beale and Keil developed their 'better-likeness' task (i.e., participants must judge which of two images is the better likeness of a particular person) specifically to bias observers, in contrast to discrimination tasks, to more holistic processing (Beale & Keil, 1995). Furthermore, in a study of face perception in children and adults, Schwarzer and colleagues found that participants' response times were significantly faster when using a holistic rather than a feature-based strategy for the same task (Schwarzer, Huber, Dümmler, 2005), indicating that processing of facial features is time-consuming. Faster response times were also found in the Feminine task of our experiment, compared to the Sex task using the same face stimuli. In sum, we suggest that in the case of the Feminine task participants' faster performance, their different pattern of fixation and their oral report about intuitive decision all indicate a holistic (i.e., low-frequency based) rather than feature-based (high-frequency) strategy.

What remains is the question about the large performance difference between both sex-related tasks. Can they be explained, as we have suggested, by more holistic versus more feature-based strategies? In the recent face perception literature, several studies have shown that the superior performance towards 'own-race' compared to 'other-race' faces that is commonly found in face identification tasks correlates with a more holistic perception of faces of ones own race, and a more feature-based processing of other-race faces (Michel, Caldara et al., 2006; Michel, Rossion et. al., 2006; Michel, Corneille, Rossion, 2007). Thus in the case of the other-race effect, holistic processing seems to be at least one factor that accounts for a difference in performance between the two tasks. We would propose a similar explanation for our results.

Furthermore, on one hand the discrimination of faces based on manipulations of their sex (a characteristic that is not expected to change over a person's lifetime) might be an extremely unnatural and therefore difficult task. The features intuitively considered as diagnostic for sex categorization that have been reported in single-face studies might not provide information adequate for the task. On the other hand, the internal representation of a face's perceived 'femaleness' might be more intuitively accessible and thus lead to more useful (possibly more low-frequency, as proposed earlier) information processing indicated by the different scanning patterns in our two tasks. We propose that these differences in face scanning might underlie the seeming contradiction in the face perception literature, between the ease with which a face is categorized as male or female, and the difficulty of a sex-discrimination task using sex-morphed stimuli. As the small number of cases in which people we know change their sex suggests, it is not surprising that discriminating faces by their sexual information is not what our visual system is tuned for.

Half-face comparison? Unexpectedly, large asymmetries were revealed in the fixation frequencies to the left and right bilateral facial features like the eyes and the cheeks (*Figure 9*). We expected observers to rather compare the left half faces to each other than the right ones, on account of the repeatedly reported left hemiface bias in face perception tasks (e.g., Butler, Gilchrist, Burt, Perret, Jones, & Harvey, 2005; Gilbert & Bakan, 1973; Mertens, Siegmund, & Grüsser, 1993). Astonishingly, our results show that participants directed more fixations to the inner half of each face (nearer to the fixation point in the middle of the screen) than to the outer half, irrespectively of the position of the face on the screen (left or right). This asymmetry was found in all tasks. This finding is surprising because it indicates that participants did not compare the corresponding features of the two faces, e.g., left eye of left face with left eye of right face. Our scanpath analyses (see *Figure 10*) support this result, showing that observers made much more frequently 'asymmetric' comparisons of the eye regions than 'symmetric' comparisons (*Figure 10*). One could assume that this behavior is an effect of the restricted time available to respond; but the results of the pilot experiment in which participants were given an unlimited time to respond revealed exactly the same pattern (results are not shown). Reduced fixation to eccentric AOIs could be an explanation, too, but the results of other studies cast doubt on such an eccentricity effect: For example, Galpin and Underwood (2005), in a comparative visual search study, had their observers make 16.5 degree eye movements without the possibility to move their head and apparently without any difficulty on their part. In our experiment, the visual angle between the left eyes of both faces was typically of about 14 degrees. Furthermore, Havard and Burton (2006) found a similar tendency to focus on the inner half of two faces in a comparison task although their faces were much smaller (about 9 by 6 degrees of visual angle) and separated by only 2 degrees. As Henderson and colleagues (Henderson et al., 2003) showed the perceptual span to be only 4 degrees around the fixation point at maximum, we can assume that our observers could not 'see' the whole facial (particularly featural) information when only fixating on the inner half. We nevertheless tested the possibility that preferred fixation on the inner facial half might be a consequence of our set-up by having another group of participants perform Experiment I while resting their chin on a soft foam ball to control only vertical head position; their head was not fixed and they were explicitly instructed not to care about moving their head a little. This was possible because the iViewX eye camera can compensate for slight head movements. However, the results (not shown in this paper) remained exactly the same: Participants fixated predominantly on the inner half of the two faces they were comparing.

The simplest explanation is that humans consider faces to be generally symmetrical, even though they actually know that this is true only to a limited extent. It intuitively appears to be a useful heuristic, and the visual system seems to keep relying on it even when task difficulty increases: our results show that when similarity between faces increased, our observers did not pay more attention to the outer halves of the face stimuli.

Observers' sex. We found surprising differences between male and female viewing strategies in both sex-related but not in the identity-related task. In the Sex and Feminine tasks, female participants looked significantly more often at the eyes, while male participants looked more often at the cheeks of the face stimuli (independent from observers' presumably more 'holistic' strategy in the Feminine task, as discussed above) . These sex-related differences in gaze behavior did not have any impact on performance, raising the question why at all male and female observers looked at the faces in a different way.

One could speculate that - in their inner representation of male and female faces - female observers place the main emphasis on the eyes (presumably including the brow region) as characteristic feature disclosing important physiognomic differences between the sexes, whereas male observers emphasize more different shape occurrences of the cheeks (and/or maybe the outer face shape at this location). These findings are in agreement with previous studies that have proposed that these two face structures code for differences in male and female faces or are used by observers for sex judgments on faces (e.g., Yamaguchi, Hirukawa, & Kanazawa, 1995, O'Toole, Deffenbacher, Valentin, McKee, Huff, Abdi, 1998).

A more technical aspect of these findings is that they point to the shortcoming of considering performance alone as a measure to investigate face processing. In our case, by relying on performance alone, we would have failed to notice sex-related differences in face processing that were revealed only in behavioral data. Obviously, it is of fundamental importance to balance observers' sex in face perception tasks (especially when sex as a facial characteristic is involved) and to analyze the data of male and female participants separately. This is even more indicated in view of the growing evidence in the recent literature for sex-related differences in performance (e.g., Lewin & Herlitz, 2002; Rehnmann & Herlitz, 2007; Guillem & Mograss, 2005; O'Toole, Peterson, Deffenbacher, 1996), behavior (e.g., the results of this study), and processing of face stimuli (in terms of brain activity, e.g., Kranz & Ishai, 2006; Guntekin & Basar, 2007).

Conclusion. Our study demonstrates the great influence of task instruction on viewing behavior and on performance when people compare faces. Using an eye-tracking technique, we could follow observer's focus of attention and thus reveal which different parts of a face were considered informative with regard to the task participants had to perform. Whilst in general confirming what has formerly been reported regarding the importance of specific facial features in single face perception tasks, our design allows us to extend these results to the more general level of identity and sex representations of faces, independent from task-related memory references. Interestingly, our data reveal that male observers look more often at the cheeks and female observers scan the eyes more when the sex of the faces is a relevant feature in the task they are performing. To our knowledge, such sex-differences in eye movements have not been reported before. Also, eye movements as well as performance differed for sex-related tasks that primarily differed in terms of instructions given to the participants. The interpretation of those findings is that observers direct their attention to different spatial scales in both tasks. Furthermore, surprising large asymmetries were found in viewing behavior as the observers' focus of attention was constantly on the inner halves of both face stimuli, suggesting that observers regarded the faces as symmetric or that at least they found this approach a useful heuristic. This finding is in contrast to the general left hemi-field bias reported for faces.

Acknowledgements

This work was supported by the Max Planck Society. We would like to thank Johannes Schultz for his support with implementing the analysis and Quoc Vuong as well as the 3 anonymous reviewers for helpful suggestions on the manuscript.

Chapter 2: Male and female faces are only perceived categorically when linked to familiar identities - and when in doubt, he is a male

Introduction

When we look at the world around us, we do not see gradual transitions between elements, be they different wavelengths of light, or different face expressions. Instead, the visual system carves our environment into separate, meaningful categories, like red or yellow colors and sad or smiling faces, via the cognitive process called categorical perception (CP). This process is fundamental to complex behavior, since it spares us from having to learn anew each time we encounter unknown objects or individuals and thus helps to reduce the overwhelming number of entities in the world to more manageable proportions (e.g., Rosch, Mervis, Gray Johnson, & Boyes-Braem, 1976; Harnad, 1987).

For the specific case of face perception, CP has been found using continua of images (morphs) created by morphing between realistic human faces of different identities (Beale & Keil 1995), expressions (Calder et al., 1996), and ethnicities (Levin & Beale, 2000). However, on the question whether the facial dimension "sex" is also perceived naturally as one of two different categories, i.e. male and female faces, conflicting psychophysical results have been reported so far (Campanella et al. 2001; Bülthoff & Newell, 2004).

Campanella and colleagues showed CP for sex (Campanella et al., 2001) using an image-morphing procedure to generate continua of face stimuli in which sex information was varied linearly between male and female faces. Additionally, however, their face stimuli were morphs between different (opposite-sex) identities. Furthermore, only few face pairs were used, and the same stimuli appeared many times per task, so that participants were being familiarized with the faces in the course of the experiment. The CP effect could thus result from categorical perception of familiar face identities rather than from CP of the sex of unfamiliar faces.

Bülthoff and Newell likewise investigated if male and female faces are discrete categories at the perceptual level and whether familiarization plays a role in the categorical perception of sex (Bülthoff and Newell, 2004). They used a 3D morphing algorithm to create artificial sex continua not only between male and female faces, but also based on single face identities that are created by changing only the sex of a face while keeping its identity constant. When using these sex continua and while increasing the number of original face identities (from 6 to 12) to reduce a potential familiarization effect, the authors could not find CP for sex. The effect only appeared when participants were either familiarized with the endpoint (i.e., most male and most female) faces of the morph continua or trained to classify all faces of the continua as male or female using a feedback procedure.

So the question whether or not there is CP for sex as a dimension of human faces remains open. The Bülthoff & Newel study suggests that processing of the sex of a face is directly linked to processing of the face's identity (as proposed before by Rossion, 2002; Ganel & Goshen-Gottstein, 2002). Yet, since it has been shown that different facial expressions and races are perceived as discrete categories (Calder et al., 1996; Levin & Beale, 2000), it seems surprising that there is no CP for such biologically and socially meaningful categories as male and female faces. As suggested by Leopold and colleagues (Leopold et al. 2006), both the time to learn as well as the storage capacity in the brain for faces can be spared by applying common transformations (changes in e.g. scale, viewing angle, expression) not to each face identity, but instead to the 'template', or reference, to which incoming face stimuli are compared. In the same vein, one can assume that the brain compares newly encountered faces to a male and female face reference, if 'male' and 'female' are discrete categories at the perceptual level. Classifying the faces of unknown individuals by their sex seems to be a prerequisite for social behaviour and communication.

Therefore, here, we revisit CP for sex using new face stimuli to deal with potential confounds that, to our opinion, might have influenced the results of earlier studies. The face stimuli in former studies were generated from 2D images

(Campanella et al., 2001) or 3D head scans of original face identities (Bülthoff and Newell, 2004); face continua were either generated by morphing one face identity with another identity of the other sex (Campanella et al., 2001; Bülthoff and Newell, 2004), or by applying the "sex vector" (calculated as the difference between the average male and the average female face of the scanned population of heads) onto each single face identity (Bülthoff and Newell, 2004). With both procedures, depending on how strongly male or female the original faces look, the continua derived from them vary in the range of "maleness" and "femaleness" they cover. Hence morph levels, as they are calculated relative to the original face of each continuum, are not comparable across continua. Even if there is a category boundary between male and female at the perceptual level, its position between the extremes would vary for each individual face morph continuum. Averaging performance in CP tasks over continua based on faces with different levels of perceived masculinity and femininity might thus cancel out any evidence for CP.

To avoid the problem of having potentially different locations of the sex boundary for each continuum, we equated the level of maleness and femaleness of all face identities by modifying the original faces before creating test continua. By using "normalized" endpoint faces, all continua should cover a similar range of maleness and femaleness and the category boundary between male and female should then be located at the same place along all face continua, with similar steps in between. We performed extensive rating experiments (as specified in the methods section) to carefully create and choose these "controlled" male and female endpoint faces. By doing this, variations of femininity and masculinity of the endpoints of each continuum and - as a consequence - variation of the location of the category boundary was kept to a minimum. An alternative to equating femininity and masculinity of the endpoint faces before creating continua would be to adjust the continua after the experiment, according to the category boundary that participants' performance reveals. By pre-equating, however, we make sure that (1) the morphing steps along the continua are of equivalent size, and that (2) for each morph level, the same number of data points is collected and entered into the analysis.

Once the "blurriness" of the location of the category boundary reduced to a minimum, our goal was to test if CP for sex does occur naturally, without reference to identity-related facial information. To this end, we conducted our study by following the classical procedure to define categorical perception, as described for example in Beale and Keil (1995), Etcoff and Magee (1992), and Bülthoff & Newell (2004). In brief, a classification task was used to locate the potential category boundary between male and female faces. A discrimination task using pairs of stimuli from different positions along the continuum was used to test if faces from one side of the boundary were indeed perceived as more similar to each other than to faces on the other side of the boundary, as expected for CP.

These two tasks were used in four different experiments. In a first "naïve" experiment, participants were tested on CP for sex without any previous knowledge or exposure to any faces. By this, we tested whether evidence for CP was obscured by the methods used in earlier studies as explained above. Three other experiments involved three different types of familiarization prior testing. The aim of these manipulations of classical CP experiments was to shed more light not only on the question whether male and female faces are naturally perceived as two separate categories, but also on the issue of the potential link between the processing of a face's sex and its identity.

According to the classical functional architecture of face processing proposed by Bruce and Young (1986), sex processing in faces is a parallel function to individual face recognition, and as a consequence, sex categorization of faces is not influenced by face familiarity. If this is true, the presence or absence of a categorical effect for the sex of faces should not depend on familiarization with the respective face identities. Levin and Beale showed in their study about CP for newly learned face identities (Levin & Beale, 2000) that short familiarization with previously unknown faces is sufficient to result in CP for continua of those identities while no CP was found for strictly unfamiliar faces. In the same logic, CP for sex should occur after familiarization with only sex-specific facial information, if the prototypical appearance of male and female faces is considered to be a priori "unknown" and the two categories have to be learned. If sex categories are, however, "linked" to individual face identities, CP for sex should only occur for familiar face identities. In the three "non-naïve" experiments, we thus tested what specific sex and identity information from faces is necessary for CP for sex to occur.

In the first non-naive experiment, participants were familiarized with the average male and female faces. By comparing the performance of naïve participants with the results of this experiment we wanted to investigate whether information about the average appearance of male and female faces, independent of identity (note that average faces are by definition devoid of facial information that is idiosyncratic to one identity but clearly belong to one sex or the other), transfers in some way to unfamiliar faces. As mentioned before, when tested on morph continua between individual familiar face identities, participants show categorical perception (e.g., Levin & Beale, 2000). Analogous to these findings, familiarization with the male and female endpoints of a sex morph continuum should lead to CP, if sex and identity are two independent dimensions of human faces.

The male and female average faces are created by morphing together all individual face identities in our database (i.e., around 200), and thus consist of information derived from faces of all perceived levels of masculinity and femininity. They might thus not represent the same "symmetric endpoints" for sex continua as the ones we obtained by the rating experiments (described in the methods section) – although this would be surprising, considering that each average was created by morphing nearly 100 faces. If we nevertheless consider the possibility that, for example, the average male face looks less strongly male than the endpoints of our continua, we might expect a potential category boundary between the two averages to be shifted away from the expected symmetric boundary between normalized male and female identities (even though the shift should be the same on all continua). Another familiarization procedure was introduced to overcome this discrepancy in the male-female range between familiarization and test stimuli, and to further examine the interplay of sex- and identity-related information in sex perception. Here, familiarization with sex information was done using male and female face identities which had the same perceived degree of maleness and femaleness than the test faces, but were not used in the following test. These faces provided participants with information about the appearance of male and female faces that also specified an individual identity. With this type of familiarization we could test whether sex-related information in faces can induce CP for sex if this information is linked to idiosyncratic facial information but has to transfer to unknown face identities.

Finally, in the last experiment, we tested whether prior knowledge about someone's identity (and sex) has an effect on the perception of the sex of face images of that person. Here, participants learned the actual endpoint faces of the sex continua that were used to test CP subsequently. This experiment served as a control, since it has been shown before that familiarization with the endpoint identities of a sex continuum leads to CP for sex.

To summarize, the aim of this study is to clarify two old but still open questions in face perception, i.e., (1) whether male and female faces are perceived categorically, and (2) how sex and identity information interact in face recognition. By investigating these issues, we intend to add evidence to the more general understanding of how human faces are perceived, processed, encoded and represented in the brain.

Methods

Three rating experiments were performed, successively, to select appropriate face stimuli for the main experiment. Design and procedure of these ratings are presented first in the order they were performed, as each was based on the results of the previous one. Methods for the main CP experiments are described subsequently.

(1) Creating Normalized Face Stimuli:

We used 3-dimensional laser scans of real heads from the database of the Max Planck Institute for Biological Cybernetics (http://faces.kyb.tuebingen.mpg.de) and the Morphable Model of Blanz and Vetter to create all face stimuli (for more details see Blanz, 2000; Blanz & Vetter, 1999; Vetter & Poggio, 1997). The general method to create sex continua based on one single identity is to calculate first a "sex vector" from the whole face database, i.e., the difference between male average and female average is calculated. Using this sex vector, an opposite-sex version of each original face can be generated. Between these endpoints (original face and opposite-sex version), morphs are created at regular intervals. Computationally, original female faces are 100 percent female and original male faces are 0 percent female. We will use this notation in percentage of femaleness (i.e., percentage of contribution of the female face identity to the morph) to describe all morphs used in this study, with 100% denoting a face derived from a scan of a female head (or a male identity feminized by 100% of the sex vector, see below). *Figure 1A* shows an example of such a one-identity morph continuum from female to male.

To choose endpoint faces equated in their perceived level of male- and femaleness for generating the sex morph continua, we conducted three consecutive rating experiments described in the following.

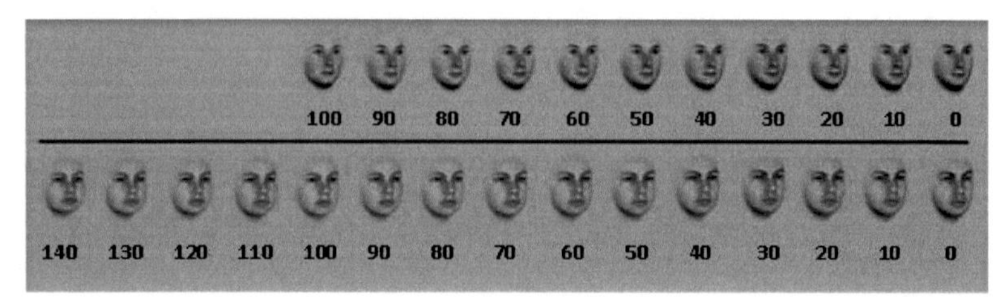

Figure 1: Sex Morph Continua. **A:** Morph continuum created by applying the sex vector on one single (here: female) face identity. Numbers indicate the percentage of contribution of the female face to create the morph. **B:** Extended morph continuum, created by morphing towards female by another 40% of the sex vector. For more details, see text.

Rating 1: Rating Original Male and Female Faces: Images of 95 male and 95 female original faces (i.e., derived from laser scans without sex manipulations) collected in the MPI face database were rated for masculinity or femininity. The hair of these faces is cropped (at the hairline) and the faces are devoid of make-up, glasses or facial hair. The faces were presented turned to the right by 20 degrees, in a 24-bit color format and on a grey background. Images subtended approximately 8 by 6 degrees of visual angle, and the average viewing distance was 57 cm. The experiments were presented on a Windows PC using the Matlab PsychToolbox (© 1984-2007 The MathWorks, Inc., Version 7.4.0) on a color monitor.

In each trial, one face was shown as long as needed for the participants to answer, there was no time pressure. Participants used the arrow keys on a keyboard to move a slider on the screen, to rate the faces on a 7-step scale, ranging from "very female" (1) to "very male" (7). They were told to answer with "ambiguous" (slider step 4) in case they could not decide if a face was that of a woman or a man. The next trial was initiated when the participant entered a response.

Eighteen paid volunteers recruited via the MPI Subject Database performed this rating experiment. Ratings (*see Figure 2A*) for female faces were distributed over the whole scaling range, while male faces were very rarely rated as female;

their ratings were mainly restricted to the male end of the rating scale. This male bias does not seem to be a phenomenon specific of our database, as other collections of face stimuli have been shown to evoke the same perception bias, even when the database consisted of silhouettes of faces in profile (e.g., Davidenko, 2007). The bias is suggested to come from the lack of additional information one is used to see in everyday life, like hair or makeup, in this kind of stimuli. While most male faces were rated as "normally male" or even "very male" (6 or 7 on the rating scale), only few female faces fell within the equivalent range ("normally female", "very female") at the female end of the scale.

Rating 2: Rating Feminized Female Faces and Original Male Faces: To obtain a sufficient number of female faces with femaleness ratings comparable to the ratings for the male faces, we feminized all 95 original female faces from our database used in Rating 1. Male faces were not modified. Each female face identity was morphed along the sex vector, away from the male end of the continuum. After some informal pilot testing, we decided to morph the female faces 40% away from their original endpoint, as we seemed to obtain ratings comparable to the ratings for male faces in Rating 1. By doing that, we generated computationally "super female" faces that are 140% female, compared to the 100% original female faces.

A group of 18 new participants were asked to rate those super female faces mixed with the unmodified male faces in the same way as described for Rating 1. As is visible in *Figure 2B*, now the ratings for the female faces are more densely grouped and shifted towards the female end of the scale, while the ratings for male faces show the same pattern as in Rating 1.

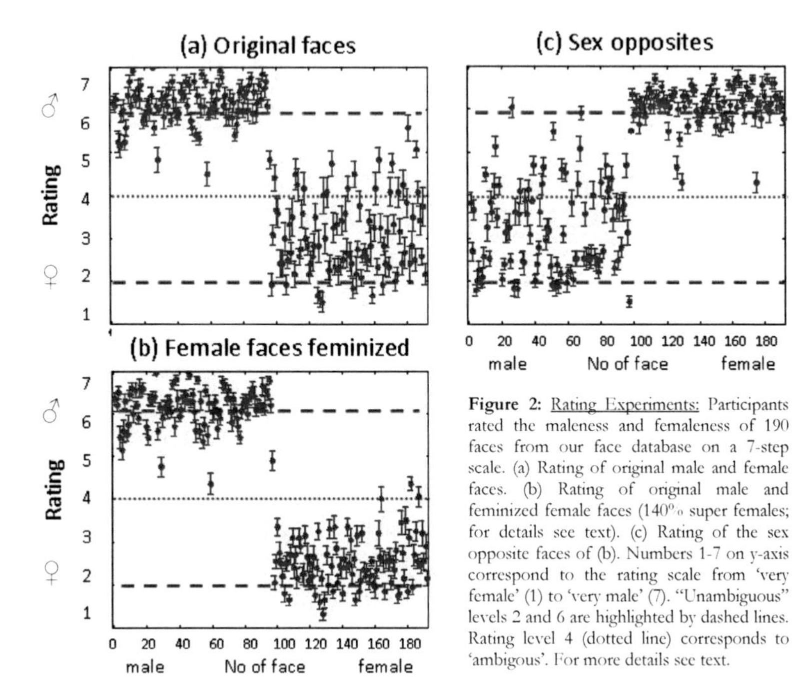

Figure 2: Rating Experiments: Participants rated the maleness and femaleness of 190 faces from our face database on a 7-step scale. (a) Rating of original male and female faces. (b) Rating of original male and feminized female faces (140% super females; for details see text). (c) Rating of the sex opposite faces of (b). Numbers 1-7 on y-axis correspond to the rating scale from 'very female' (1) to 'very male' (7). "Unambiguous" levels 2 and 6 are highlighted by dashed lines. Rating level 4 (dotted line) corresponds to 'ambigous'. For more details see text.

Rating 3: Rating Opposite-sex Versions: For the CP experiments, we wanted symmetrical face continua. This means that every endpoint identity used to create a single-identity sex morph continuum has to be perceived as "normally male" or "very male" on one end of the continuum and "normally female" or "very female"" on the other end. Therefore, we also acquired ratings for the maleness or femaleness of the other-sex version of every face identity (so far, ratings were only obtained for faces with their original sex). Here, male version (0% female) of original female faces (140% female) were rated, and female versions (140%) of original male faces.

Twenty participants rated these male (i.e., modified females) and female (i.e., modified males) faces as described in Rating 1. Results are shown in *Figure 2C*. The pattern is the same as in Rating 1: Ratings for male faces are mainly restricted to the male end of the scale, while female faces are often classified as "ambiguous" or even male.

Choosing 'Symmetrical' Stimuli: The rating scale consists of 3 "zones": female (1-3), ambiguous (4) and male (5-7) ratings. The middle ratings 2 for female and 6 for male faces were defined as "unambiguous" values of maleness and femaleness. The mean ratings for each face (for the original male or the feminized female face and their other-sex versions, respectively) were thus compared to the values 2 or 6 using a 2-sided t-test. All faces with mean ratings significantly different from these values were excluded. Among the remaining faces, there were identities that had both their female and male version within range of the unambiguous values for both sexes. We chose 20 of those faces (10 male and 10 female original identities) to create symmetrical sex continua with endpoint faces within the desired range of maleness and femaleness for the main experiment.

Sex-specific Sex Perception: Since differences between male and female observers have been reported in the face perception literature, we compared the rating results of male and female participants, but found no significant effect of observers' sex in any of the ratings.

(2) Main Experiment: CP for the Sex of Faces

In the following, we first describe methods and procedures common to all four CP experiments; particular details of each procedure are then stated in separate sections.

Stimuli

Male and female versions of ten male and ten female face identities selected through the rating procedures described above were used as endpoint faces to generate one-identity sex morph continua. All endpoint faces had been rated as "unambiguously" male and female. Between these endpoints, thirteen equally spaced morph faces were created (i.e., including the original female face at 100%). Of each face identity, there were thus fifteen stimuli, ranging from male (0%) to female (140%), and our continua covered a range of 140% morph distance altogether (instead of 100% as in classical sex continua). See *Figure 1B* for an example of such an "extended" sex morph continuum.

Design *(a) Discrimination Task*

Previous studies in our lab with one-identity sex continua have revealed that participants are performing at around chance level in a classical XAB match-to-sample paradigm even when the two face images are 30% apart in morph distance (Bülthoff & Newell, 2004). Increasing the morph distance to make the task easier decreases the number of possible face pair combinations within one continuum, which makes it harder to measure a peak in performance and thus show evidence for CP. Yet, since we extended the continua by four morphs, we could afford to use pairs of faces at a distance of 40%, to make sure participants were able to accomplish the task.

Moreover, two other modifications of the classical CP paradigm were made: (1) Instead of the usual XAB task we used a simultaneous same-different task, to avoid the memory load that the match-to-sample paradigm requires (see Calder et al., 1996). By doing this, we made sure that our discrimination data does not reflect short-term memory capacity, but a perceptual phenomenon. (2) Earlier studies in our lab have shown that a same-different task with faces manipulated along the sex dimension only is somehow unintuitive to participants and might lead to image matching rather than a sex judgment. Therefore, here, we used a more accessible and, as we found earlier, easier task (Armann & Bülthoff, 2009): Our participants had to answer, for every pair of stimuli, whether one of the two faces was more feminine or masculine than the other, or whether they were exactly identical.

Newly learned faces or familiarization with face stimuli in the course of an experiment lead to categorical perception (e.g., Viviani et al., 2007). We thus tried to keep exposure to each of the identities used to create our sex continua to a minimum. Of each one-identity continuum, every participant saw only some of all possible pair combinations. Given that there are fifteen face stimuli in each continuum, and pairs were shown at a morph distance of 40%, there were eleven possible "different pairs" for each identity (0-40, 10-50, 20-60, 30-70, 40-80, 50-90, 60-100, 70-110, 80-120, 90-130, 100-140). Five of these pairs were pseudo-randomly chosen for each identity as "different pairs", and five morph images of the remaining pairs were chosen for "same pairs", counterbalanced across participants. Every participant was shown ten "different pairs" at every possible morph level combination along the continuum (while identities varied across morph levels and participants), and only these were entered into the analysis. Together with the "same pairs", there were 210 trials in total (110 "different" trials).

A trial consisted of two face images from the same continuum shown next to each other on the screen, separated by approximately 8cm. Participants were asked to decide for each face pair whether one of the faces was more male or female than the other or if they were exactly identical, as fast and as accurately as possible. Participants pressed one button on the keyboard to answer that there was a difference between the two faces (no matter which one of the faces was more male/female), and another one to answer that they were exactly the same. Response buttons were counterbalanced across participants.

Each trial started with a 500ms fixation cross. The face images then appeared and remained on the screen for 2500ms. Participants could only respond during that time; otherwise no response was recorded and the experiment continued.

They were informed about this time constraint and could experience it in a training phase preceding the experimental block. An inter-trial interval of 1000ms followed a participant's response before the start of the next trial. Trial order was randomized. The training phase consisted of 10 trials with feedback. The training face pairs (shown in random order) covered the range of the sex continua and the training identities were not used again in the experiment.

Design *(b) Classification task*

A classification task consisting of a two-alternative forced-choice (2AFC) paradigm immediately followed the discrimination task. All images from all continua were used in the classification task, yielding a total of 300 trials (20 identities x 15 face images). Participants were shown one face per trial and asked to decide as fast as possible if it was male or female, by pressing the key w and z on the keyboard (response buttons for "male" and "female" responses counterbalanced across participants). Each face was preceded by a 500ms fixation cross. The face image remained on the screen for 1500ms. Participants could only respond during that time; otherwise no response was recorded and the experiment continued. An inter-trial interval of 1000ms followed a participant's response (or the end of the presentation time) before the start of the next trial. Trial order was randomized.

Procedure

The experiments were presented on a Windows PC using the Matlab PsychToolbox (© 1984-2007 The MathWorks, Inc., Version 7.4.0) on a color monitor. Distance to the monitor was about 57 cm, and each stimulus face measured approximately 8 by 6 degrees of visual angle. All participants, apart from those in the first (naïve) experiment started with a familiarization procedure which is described in the respective experimental sections below. All participants performed the same-different task first, followed by the classification task. Participants received self-timed breaks after every 100 trials in both tasks. Instruction was written on the screen at the beginning of a task and the experimenter was present at that time for answering potential questions.

Data Analysis

Since all continua have been controlled for ranging between endpoints of similar perceived masculinity and femininity, we analyzed and present data averaged across all participants and face identities. Percent of correct "different" judgments for "different trials" are presented for the same-different task, and percent of "male" judgments for all trials for the classification task.

Referring to classical studies on CP (e.g., Beale & Keil, 1995; Levin and Beale 2000) our approach was twofold: (i) From participants' responses in the classification task, the category boundary is detected from the expected shift in judgment from one category to the other. We determined the face pair straddling the male-female boundary, i.e., pairs including one individual face classified as male on more than 66% of trials and another classified as female on more than 66% of trials. (ii) We then tested for significant categorical perception effects by entering the accuracy data from the discrimination task into a Repeated Measures ANOVA, using a deviation contrast to compare accuracy on cross-boundary pairs to the mean accuracy on all pairs combined. Increased discrimination accuracy on the cross-boundary pair was taken as an indication of CP.

Experiment 1: Naïve participants (« naïve » condition)

Seventeen participants (8 male) accomplished the two tasks assessing CP as described above. All participants were naïve as to the purpose of the experiments and unfamiliar with the faces of the MPI face database. The total duration of the experiment was about 1h.

Experiment 2: Familiarization with the average faces ("averages" condition)

Participants were familiarized with the male and the female average faces before the main experiment. They were shown images of both faces from different viewpoints and in different sizes, and with a written name to emphasize the faces' sex. The male average face was called "John", the female average "Lisa". Along with each face image, participants were asked to answer a question regarding a character trait (e.g., "how intelligent is Lisa?", "how happy do you think John is?"), using a 7-step scale on the screen and the arrow keys on the keyboard. The questions were randomly chosen from a list of 46 different character traits. Time to look at the faces and to respond was not restricted, and participants were not aware of following experiments. The task consisted of a total of 50 trials (25 trials per average face, respectively). After familiarization, participants performed an old/new recognition task with the two average faces presented in random order, intermixed with 32 distracter faces from our database that were not used in subsequent experiments.

Twenty-two participants (11 male) went through the familiarization procedure and subsequently performed both CP tasks. All of them could correctly identify the average faces after familiarization and were naïve as to the purpose of the familiarization and the following tasks. The total duration of the experiment (including the same-different and the classification task) was about 1h10.

Experiment 3: familiarization with other faces of the same sex range ("otherIDs")

As in the "averages" training procedure (see Experiment 2), each of the four "other identity" faces was shown in different sizes and from different viewpoints and each was associated with a sex-specific name tag ("John", "Thomas", "Lisa", "Mary"). Along with the face images, in each trial, participants were asked questions regarding character traits, as in Experiment 2, and had to answer using a slider and a 7-step scale on the screen. The training procedure consisted of 80 trials (20 trials per face identity). Images of all four different identities were intermixed and shown in randomized order. After familiarization, participants performed an old/new recognition task with the four learned faces presented in random order, intermixed with 32 distracter faces from our database that were not used in subsequent experiments.

Nineteen participants (8 male) completed this experiment. All of them could correctly identify the four learned faces in the old/new task and were naïve as to the purpose of the familiarization and the following tasks. Total duration of the experiment (including the same-different and the classification task) was about 1h15.

Experiment 4: Familiarization with the endpoint faces ("testfaces" condition)

Here, participants were familiarized with all endpoint faces (i.e., the originals and opposite-sex versions) that were used in the subsequent CP tasks. To cope with the much higher number of faces to remember, the face images were shown without names, but the pronouns "he" and "she" in the questions on the screen clearly pointed to the sex of each face. Each face identity was shown four times as female and four times as male version, yielding 160 trials altogether. Images of all identities were intermixed and shown in randomized order. Familiarization was done as in Experiment 3: For each face, participants had to answer a question regarding a character trait, using a slider and a 7-step scale on the screen. Subsequently, participants were asked to identify the faces they had seen before, intermixed with the 32 distracter faces used in Experiment 3.

Eighteen participants (9 male) performed this experiment, all of them naïve as to the purpose of the familiarization and subsequent experiments. Four participants (2 male) were removed from the analysis, because their identification performance of the learned faces was below 75% correct or because they did not complete both tasks. The familiarization procedure was followed by the same-different and the classification task. Total duration of the experiments was about 1h30.

Results

The four graphs in *Figure 3A* show the results of the classification task – averaged across all participants – for each experimental condition to allow better comparison. The ordinate represents the percentage of "male" responses for all trials, the abscissa the morph levels along the sex morph continuum. Classification data was fitted using the `psignifit` toolbox version 2.5.6 for Matlab (see http://bootstrap-software.org/psignifit/) which implements the maximum-likelihood method described by Wichmann and Hill (2001a). To compare classification data across conditions, the inflection point (the point of subjective equality, PSE, at 50% performance) and the smallest morph difference participants were able to discriminate (the just noticeable difference, JND) where calculated for each curve.

Figure 3B shows, below each classification curve, the results of the same-different task from the same condition. The ordinate shows percent of correct "different" judgments (only different trials were analyzed). Along the abscissa, pairs are defined as follows: *[morph level left face + morph level right face, divided by 20]*. Pair 2, for example, defines a face at morph level 0%, shown alongside a face at morph level 40%; pair 5 defines a pair including a face at level 40% and a face at level 60%.

Category Boundary:

Classically, the location of the category boundary is predicted from participants' judgments in the classification task, i.e., it would be expected at the point along the morph continuum where perception of the sex of the face stimuli changes abruptly from male to female.

Cursory visual inspection of the classification curves does not reveal s sharp step between male and female categories. We used the classical method used for example by Beale & Keil (1994) to predict performance in the same-different task. To that end, we defined 33% and 66% cut-offs for the category boundary, where one face morph was classified as female in 33% of the cases and as male in 66% of the cases, and another one as female in 66% and as male in 33% of the cases (see *Figure 3A,* solid lines). If stimuli along a continuum are perceived categorically, a peak in accuracy would be expected in the same different task for the pair that straddles the boundary. The pair that (approximately) corresponds to the location of these cut-offs, is pair 9 (including morph levels 70 and 110) in the averages and the otherIDs condition, and pair 8 (including morph levels 60 and 100) in the naïve and the testfaces condition. These pairs are highlighted by dotted vertical lines in *Figure 3B*. Since crossing the category boundary should cause an increase in accuracy, we would also expect pairs to either side of the boundary, i.e., pairs that might only partially straddle the boundary, to be more easily discriminable than fully within-category pairs.

Peak in Performance:

To test our predictions, we entered the data from each same-different task into a Repeated Measures ANOVA, using a deviation contrast to compare the performance on every pair against the grand mean of all pairs. In the "naïve" condition, performance is slightly higher around the very center of the continuum (however, not exactly the predicted boundary), but the increase is not significant. In the averages condition, although overall performance is better than in the naïve condition, and although there seems to be a peak at pair 9 (the predicted cross-boundary pair), the ANOVA does not reveal any significant differences between pairs. Performance data in the otherIDs condition shows slightly higher accuracy around the predicted location of the category boundary. The effect is however only significant for pair 7 [$F(1,18) = 10.496, p < 0.005$], i.e., a pair consisting of a face at 50% and one at 90% morph level. The peak in performance would be expected around the 90% morph level. The discrimination data from the testfaces condition shows a clear peak in performance, with pair 7, [$F(1,17) = 10.961, p = 0.004$], 6 [$F(1,17) = 15.814, p = 0.001$], and 5 [$F(1,17) = 7.521, p = 0.014$] each being discriminated more easily than the mean of all pairs along the continuum. While this peak is also slightly shifted away from the predicted location towards the male end of the continuum, it is clear evidence for better discrimination between pairs that (at least partly) straddle the boundary between male and female, compared to pairs closer to both ends of the sex continuum.

Comparison across Conditions:

Figure 4 shows several measures from both tasks to compare performance directly across conditions. One-sample t-tests reveal that the PSEs of all curves differ significantly from 70%, i.e., the actual center of the morph continua *[naïve: t(16) = -.3.348, p<0.006; averages: t(21)=-6.969, p<0.001; otherIDs: t(18)=-5.386, p<0.001; testfaces: t(17)=-4.636, p<0.001]*. Note that, since our morph continua have been chosen for their symmetrical perceptual range from male to female, the PSE would be expected to coincide with the center at 70% (as each continuum ranges from 0% to 140%).

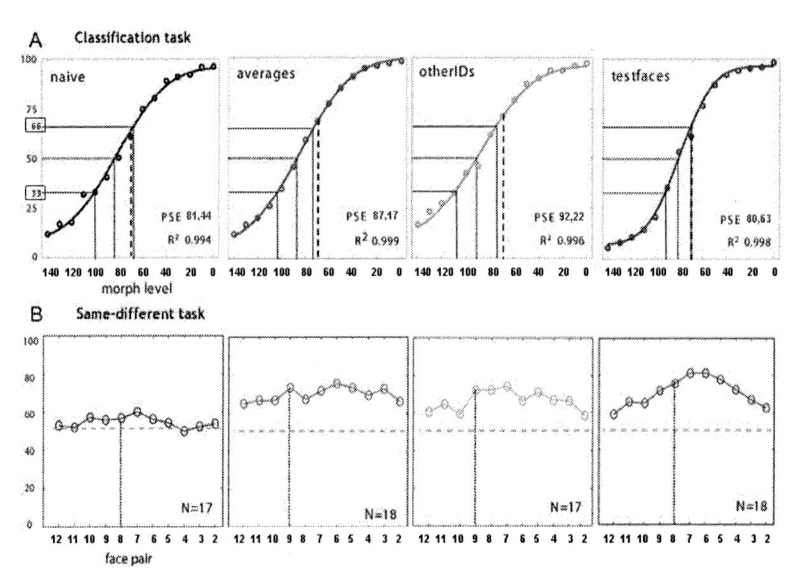

Figure 3. Results from the CP experiments. **A**: Classification data, fitted using the Matlab toolbox psignifit. The x-axis indicates percentage of classification as male, the y-axis indicates morph level. Dotted lines indicate the PSE, dashed lines the center of the morph continuum at 70% morph level. Category cut-offs are indicated at 33% and 66% male classification (see text for details). **B**: Percentage of correct responses for "different trials", from the same-different tasks. Horizontal dashed lines indicate chance level, dotted vertical lines the approximate location of the cross-category pair, determined from the classification data. Face pairs are defined as follows: *[(morph level left face + morph level right face) / 20]*. Pair 2, for example defines a face at morph level 0%, shown alongside a face at morph level 40%; pair 5 defines a pair including a face at level 40% and a face at level 60%.

The testfaces curve has the PSE value closest to 70% (59.30, SE=2.31), second comes the PSE of the naïve (58.98, SE=3.28) curve. The PSE values of both other curves (averages: 52.37, SE=2.53; otherIDs: 48.87, SE=3.92) depart more strongly away from the middle of the continua, towards the female end (i.e., above 70% morph level, see *Figure 4*). This shift on all curves indicates that morphs on the female side of the continuum are judged as male, even though the endpoint faces of the continua have been chosen as being symmetrically male and female based on the rating experiments.

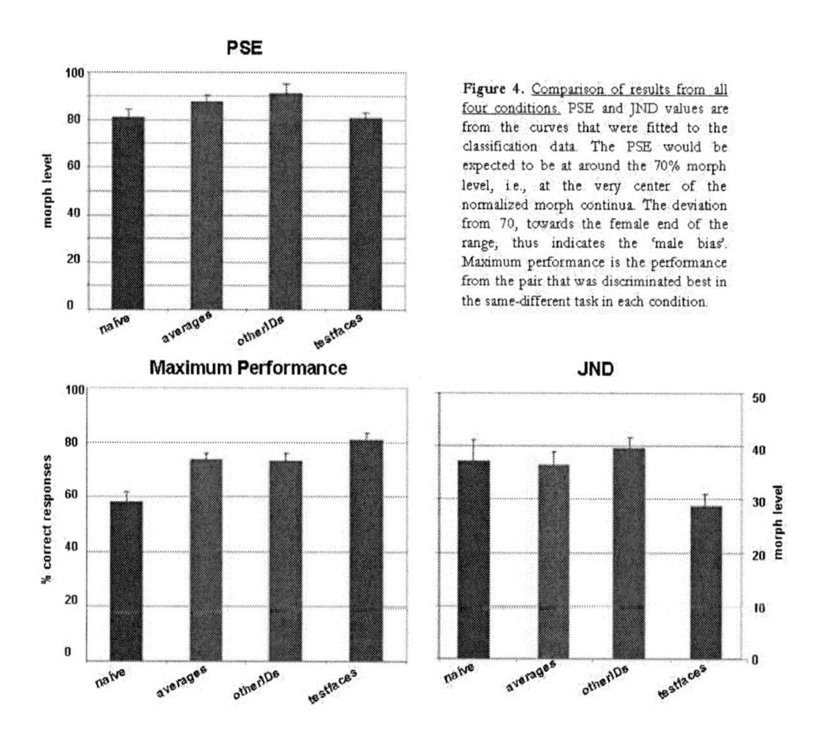

Figure 4. Comparison of results from all four conditions. PSE and JND values are from the curves that were fitted to the classification data. The PSE would be expected to be at around the 70% morph level, i.e., at the very center of the normalized morph continua. The deviation from 70, towards the female end of the range, thus indicates the 'male bias'. Maximum performance is the performance from the pair that was discriminated best in the same-different task in each condition.

A One-way ANOVA with condition (naïve, averages, otherIDs, testfaces) as between-factor *[F(3,72) = 3.414, p = 0.22]* and Bonferroni-corrected pairwise post-hoc tests show that the "just noticeable difference" (JND) of the naïve, averages and otherIDs curves do not differ from each other *[all $p > 0.05$]*, while the testfaces curve has a significantly lower JND *[naïve vs. testfaces: $p = 0.003$; averages vs. testfaces: $p = 0.039$; otherIDs vs. testfaces: $p = 0.035$]*. This is in line with the overall shape of the curves: As mentioned before, only the testfaces curve resembles a step-like function, indicating two categories with a "switch" in between, while the three other curves indicate a rather linear change in perception from female to male.

It is visible from *Figure 3B* that performance is better in the three conditions where participants had to go through a familiarization procedure before testing. If we compare the performance on the pair with the highest accuracy from every condition ("maximum performance") in *Figure 4,* A One-way ANOVA with condition (naïve, averages, otherIDs, testfaces) as between-factor *[$F(3,72) = 10.007$, $p < 0.001$]* and Bonferroni-corrected pairwise post-hoc tests reveal that this difference between the maximum performance in the naïve condition and the maximum performance in all other conditions is significant *[all $p < 0.03$]*. Maximum performance in the averages, otherIDs and testfaces condition do not differ significantly from each other.

Discussion

The main purpose of this study was to create optimal experimental conditions and stimuli for finding categorical perception of the sex of human faces. Unfamiliar faces were manipulated in their sexual appearance, while individual identity information was kept constant. Additionally, the degree of perceived maleness and femaleness of the endpoint faces used to create continua was strictly controlled. Nevertheless, we found no evidence for naturally occurring CP for the sex of faces. This is in accordance with the study by Bülthoff and Newell (2004), where the authors used the same face database (although different degrees of male- and femaleness were not taken into account), and found no CP for sex when the identity of the faces was kept constant while only sex information was morphed between male and female. These findings raise the question what, if not sex information alone, triggers the emergence of CP for male and female faces, as it has been reported in earlier studies (e.g., Campanella et al., 2000). To test whether the effect was rather due to familiarization with the test face identities in the course of the experiment than to categorical perception of the sex of unfamiliar faces, we carried out the three modified CP experiments that included a familiarization phase before the discrimination and classification tests: Familiarization was done with the average faces, additional male and female faces, and the test faces themselves.

When familiarized with sex-related information about the "typical" appearance of male and female faces provided by the average faces (which are suggested to represent the "norms" used to encode male and female faces; e.g., Jaquet & Rhodes, 2008) or by other individual face identities, participants performed better overall in the same-different task, compared to naïve participants *(see Figure 3B)*. This is not surprising, as during the familiarization process they get used to the characteristics of faces from our database as well as to the range of stimuli. However, despite higher performance than in the naïve condition, participants did still not show a "step" in classification responses, indicating the existence of a category boundary between male and female *(see Figure 3A)*. The fact that the JND *(Figure 4)* of the naïve, averages and also the otherIDs curves are all the same also indicates that familiarization with average faces does not lead to categorical perception of male and female faces, and neither does familiarization with other individual identities. There was also no significant peak at the corresponding pairs in discrimination performance along the morph continuum in the same-difference tasks of the naïve, averages and otherIDs conditions *(Figure 3B)*.

Familiarization with a male and female average face provides participants with information about the average appearance of how faces of both sexes look like, independently of traits that are characteristic to the face of a specific person. One might think, since these faces do not represent real people, that sex-related information from the average faces is not as "accessible" to the observer as information from individual face identities. If, however, sex and identity are both independent dimensions in face space, then training with both category prototypes on the dimension sex (male and female) should lead to CP, analogous to CP for a continuum of two familiar face identities (e.g., Beale & Keil, 1995). The present study shows that familiarization with the averages or "norms" for male and female faces does not result in CP along the sex continuum. Our results rather suggest that the perception of somebody's sex is not independent of the perception of the same person's identity. More evidence for this interpretation comes from the performance data for the otherIDs condition in this study: Here, familiarization is done with individual faces, hence appearance of male and female faces (and the difference between the two) is linked to idiosyncratic character traits. When tested on new unfamiliar faces, however, participants do not show clear evidence for CP, indicating that the sex information, although learned, cannot be extracted or separated from the corresponding identities and transferred to new faces.

One could argue that, unlike the otherIDs faces or the testfaces themselves, the male and female averages do not represent the actual end of the sex morph continua from female to male. They were created by averaging all 95 male and 95 female faces and thus indicate the central tendency of male and female characteristics of human faces. Since the endpoint faces for the sex continua were chosen based on ratings that judge them as "normally male" and "normally female", we can suppose that the averages would fall into a similar range on the male – female rating scale. Even if we assume that this is not the case, then we would still expect CP for male and female faces after familiarization with the averages. The category boundary might then simply not lie at the center, but at a slightly different location along the symmetrical sex continua used for testing. However, this is only possible if sex-specifying information is abstracted from specific faces and transferred to unfamiliar identities. Yet, as discussed above, there is no sign for CP emerging after familiarization with the male and female averages, be it at the expected or at any other location.

Only participants who were familiarized with the identities of the test faces show a clear discrimination performance pattern that would be expected for categorical perception, as well as a steeper classification curve (*Figure 3*). The latter is in line with a significantly smaller JND of the testfaces classification curve compared to the other three conditions (*see Figure 4*). The accuracy data from the same-different task of the testfaces condition reveals a significant peak around the predicted level, although shifted a little bit towards the male end of the continuum. This slight discrepancy has been reported before in CP studies and might just be due to the fact that the discrimination task evaluates an ability that is more related to a perceptual state, while the classification task taps more into cognitive processes (e.g., Sigala et al., submitted). Note that participants in the testfaces condition had to memorize a much higher number (20 faces in 30min max) of face identities than participants in the averages (2 faces in 10min max) and the otherIDs (4 faces in 15min max) conditions, while they had less time for every identity. Thus even the shorter exposure and probably less good encoding of each single identity leads to a considerable difference in discrimination and classification, compared to the other conditions, as expected for CP.

To summarize, our results indicate that sex and identity information in faces is not processed in parallel, as was suggested in the classical Bruce & Young model of face perception (Bruce & Young, 1986). Rather, the perception of the sex of a face seems closely linked to the perception of its identity, as stated in the single-route hypothesis. This hypothesis is based on earlier findings that show, for example, that participants in a classification task could not selectively attend to either sex or identity without being influenced by the other, irrelevant, dimension

(Ganel & Goshen-Gottstein, 2002). In another study, participants were quicker at judging the sex of familiar faces compared to unfamiliar ones, indicating that identity perception influences the perception of the sex of faces (Rossion, 2002).

The influence of face familiarity on sex categorization might seem surprising, as one of the first things we can say about a person we do not know is whether it is a man or a woman. This "categorization" of just everybody in our environment into male or female is also of biological and social significance. Moreover, in high-level adaptation studies, it has been shown that face identities are encoded relative to sex-specific rather than relative to a generic norm (Jaquet & Rhodes, 2008), suggesting that the brain indeed stores a general representation of what is male and female in faces. However, it seems that this representation is not independent from a representation of every idiosyncratic face identity and that information regarding the two traits is processed, at least partly, in an interconnected way.

Brain imaging data shows that it is surprisingly difficult to find sex-specific neural responses to faces: The responses are weak and widely distributed across the whole face network (Kaul, Rees & Ishai, 2010). A possible interpretation of these results, in the light of the present study, is that not only some specific neurons, but most face-selective neurons in the brain do have information about the sex of the face stimulus they are responding to, and that this information is not specifically extracted and gathered separately from the main character trait of each face, i.e., its individual identity.

Our morph continua have been normalized prior to the CP experiment and thus all cover the same symmetrical range from male to female. Variations in the location of the category boundary or - if there is none - the point of subjective equality (PSE) on every continuum and for every participant should thus be minimized. Surprisingly, we still observed a very consistent male bias, i.e., a deviation of the PSE of the classification curves from the actual center of the morph continuum at 70% (*see Figure 4*), towards the female end of the continuum, in every condition. This bias reveals that participants are more likely to say that a face is male than female when in doubt, and even when the stimuli are taken from within the female side of the continuum (i.e., where participants should not have any doubt at all). This phenomenon could of course be a sign that our morph algorithm is manipulating the face stimuli in a non-linear way. In this case, a bias observed in the PSE might just reflect the non-linearity of the continua that were created between the endpoint faces. However, we choose unambiguously male and female faces for the endpoints of the continua, based on the results from the ratings. It is important to note that these endpoint faces were not manipulated anymore after the rating experiments. They at least should thus be classified as clearly male and female in the main experiment, since this was our selection criteria. A look at *Figure 3A,* however, reveals that even the female endpoint faces, i.e., the faces at 140% morph level in every graph, are not classified as female all the time; in the naïve, the averages and the otherID condition, even the most female faces are rated as female in less than 88% of the cases (naïve: 87.65, SE=4.28: averages: 87.86, SE=2.28), even in only 83% of the cases in the otherIDs condition (SE=2.57). Only in the testfaces condition, female judgments for the most female faces reach 95% (SE=1.69). Independent samples t-tests show that the difference in these judgments for female endpoint faces between the testfaces condition and all other conditions is significant *[all $p < 0.04$].* Male judgments at the male end of the continuum (at 0% morph level) are always above 96% and do not differ between conditions.

Interestingly, Troje and colleagues (Troje et al., 2006), when investigating adaptation aftereffects in the perception of the sex of male and female point-light walkers, made the same rather unexpected discovery: A bias toward seeing more male than female walkers in a set of stimuli before adaptation, even though they then choose their stimuli to control for this bias and thus keep the "sex range" of the walkers symmetrical, remained almost constant.

These findings suggest that there is a perceptual or cognitive bias to answer "male" when in doubt about a person's sex. In the case of human faces this phenomenon has been suggested to result from an anatomical lack of distinctly female features (e.g., Enlow, 1990) in faces in general. Answering "female" while classifying a face's sex would thus be a "no male traits" response. Apart from the fact that body shape definitely plays a role in defining somebody's sex, this bias could imply that external features, i.e., hairstyle, makeup, clothing, maybe even behavior, might be used as cues to a person's sex, more than just physical appearance of the face itself. Interestingly, all these external features are defined and shaped by culture and thus are not biologically "hardwired" and universal. It could also be that misclassifying a male person as female has generally proved to be potentially more dangerous than misclassifying a woman as a man in the history of humans. We do not want to overspeculate here; however, our results, together with results from other current studies (e.g., Troje et al., 2006) suggest that there might be more to this male bias, in addition to the stimulus-driven bias that is usually found in face stimuli when they are deprived of external information.

Acknowledgements

We want to thank Mario Kleiner for assistance with the Face Database and Johannes Schultz for help with data analysis. This research was supported financially by the Max Planck Society.

Chapter 3: Race-specific norms for coding face identity and a functional role for norms

Introduction

Humans recognize faces with remarkable accuracy, despite the fact that faces all share the same basic structure and differ only in subtle aspects of featural and configural information. This ability is thought to rely on adaptive face-coding mechanisms that are dynamically updated in response to changes in the statistics of faces encountered over time (for recent reviews see Rhodes & Leopold, 2011; Clifford & Rhodes 2005). By representing every face relative to a stored average or prototype that functions as a norm, such "norm-based coding" may allow the visual system to see past the shared, highly redundant, structure of faces and to focus on what is distinctive about each individual.

Support for a norm-based representational framework for faces comes from different lines of research. First, it seems that people spontaneously abstract averages or prototypes from sets of seen faces (Bruce, Doyle, Dench, & Burton, 1991; Walton & Bower, 1993; Inn, Walden, & Solso, 1993), i.e., the face corresponding to the central tendency of a series of faces seems familiar, even when this average or prototype has not been seen. Furthermore, it has been found that distinctive faces that lie further from the average in a face space framework are recognized better than typical ones (Valentine, 1991; 2001), and that caricaturing a face, i.e., exaggerating how it differs from the average, can facilitate its recognition (Benson & Perret, 1994; Byatt & Rhodes, 1998; Calder, Young, Benson & Perret, 1996; Lee, Byatt, & Rhodes, 2000; Rhodes, 1996; Rhodes, Brennan, & Carey, 1987). Recent neuroimaging and neurophysiological findings are also consistent with norm-based coding of faces: With increasing distance from the average, faces elicit stronger fMRI activation in the human fusiform face area (Loffler, Yourganov, Wilkinson, & Wilson, 2005) and increase firing rates of face-selective neurons in monkey anterior inferotemporal cortex (Leopold, Bondar, & Giese, 2006), whereas responses to average faces are weaker.

Recently, studies of face-identity aftereffects have provided the most direct evidence that identity is coded relative to a norm. Adapting to a face biases perception towards the "opposite" identity in face space relative to the average face (Leopold, O'Toole, Vetter & Blanz, 2001; Leopold, Rhodes, Müller, & Jeffery, 2005; Rhodes & Jeffery, 2006). The fact that perception is biased in a selective way towards the opposite identity relative to the average, not just unspecifically away from the adapting face, strongly supports the notion that faces are coded as deviation vectors from a norm (in terms of distance and direction), and that some sort of opponent coding mechanism is involved (Rhodes & Jeffery, 2005; Tsao & Freiwald, 2006). Norm-based face coding appears to be implemented by a two-pool opponent coding mechanism at the neural level, where all possible values along a single dimension in face space are coded by the relative output of only two oppositely tuned pools of neurons. Adaptation to the average face itself, for example, does not shift the perception of non-average faces (Leopold, O'Toole, Vetter, & Blanz, 2001; Webster & MacLin, 1999), in accordance with an opponent coding model where the average is perceived when both pools of neurons produce equal output strength. Also, aftereffects become larger for adaptors with more extreme distortion levels relative to the norm (Robbins, McKone, & Edwards, 2007), as predicted by opponent-coding: Here, because adaptation reduces responses in proportion to the initial unadapted firing rate, adaptors furthest from the norm produce the maximum change in the response ratio of the two pools (e.g., Maddess, McCourt, Blakeslee, & Cunningham, 1988; Movshon & Lennie, 1979).

The concept of a norm-based face space raises the question of whether there really is one central norm representing all dimensions on which faces can differ, or whether there are multiple norms corresponding to a variety of distinct face categories. For the sex and race of faces, "category-contingent" aftereffects have been found, where opposite figural (distortion) aftereffects are generated simultaneously for different groups of faces. Adaptation to "contracted" male and "expanded" female faces, for example, led to opposite aftereffects in both face categories, respectively, with undistorted male faces looking slightly expanded and undistorted female faces looking slightly contracted (Little, DeBruine, & Jones, 2005; Bestelmeyer, Jones, & DeBruine, 2008; Jaquet & Rhodes, 2008). The same opposite distortion aftereffects can be induced with Asian and Caucasian faces (Jaquet, Rhodes, & Hayward, 2007). The most normal-looking face configuration shifts towards the adapting distortion, consistent with a shift in the norm. These findings have been interpreted as evidence for category-specific norms for faces of different races and sexes. Category-contingent aftereffects are, however, small, and figural aftereffects induced by viewing distorted faces of one category sometimes transfer strongly to faces of another category (Jaquet & Rhodes, 2008; Jaquet, Rhodes, & Hayward, 2008). This transfer presumably reflects the fact that male and female faces, as well as faces of all races, share the same basic structure. Therefore, identity could in principle be coded using a generic norm, independent of the sex or race of a particular face.

Faces of different races represent distinct visual as well as social categories. Current face space models propose that own- and other-race faces are represented in different locations of face space (Valentine, 1991; Valentine & Endo, 1992). In these models, same-race faces are located around the average face of the space, while other-race faces form a dense cluster far away from the average. Such a representation takes into account that faces of one's own race are usually more easily recognized than faces of a more unfamiliar race (the so-called "other-race effect", e.g., Meissner & Brigham, 2001; Michel, Caldara, & Rossion, 2006; Michel, Corneille, & Rossion, 2007). In the present study, we determined whether other-race faces are consequently encoded around a norm of their own, rather than around a generic norm for the whole face space. To that end, we tested whether face identity aftereffects (e.g. Leopold, O'Toole, Vetter, & Blanz 2001; Rhodes & Jeffery, 2006) were larger when adapt-test face pairs came from trajectories made relative to a race-specific average versus a generic (mixed race) average. Adapt-test pairs that lie opposite each other (at the same distance) in face space generate larger identity aftereffects than non-opposite test pairs (Rhodes & Jeffery, 2006). Therefore, comparing the size of aftereffects on a trajectory created by morphing through the same-race average face with aftereffects on a trajectory created by morphing through a generic average should allow us to identify the "truly" opposite adapt and test faces, i.e., the trajectory that generates the largest aftereffects. The average used to create this trajectory would represent the center of the observers' mental face-space representation and therefore the true psychological norm.

On each trial, participants were first exposed to an "antiface" (i.e., an identity lying opposite to its respective target identity in face space) and subsequently had to identify target faces at various identity strengths. To generate the antifaces, Asian and Caucasian male target faces were morphed along a trajectory through a "reference face" that could be either an average or a control face (see below), and further away from it on the opposite side. We used different reference faces, to represent different possible norms, and thus create different trajectories and antifaces. For examples of trajectories and reference faces, see *Figure 1*. The reference was either (1) created from morphing together Asian or Caucasian faces, respectively, i.e., a race-specific, or "same-race" average, (2) a combined average from Asian and Caucasian faces, i.e., "generic", or "mixed", or (3) another face identity of the same race as the respective target faces, i.e., Asian for the Asian target faces and Caucasian for the Caucasian targets. This identity was taken from a location in face space that does not correspond to the center, and is called the "non-central" reference throughout this manuscript. These five reference faces that were used to create trajectories are shown in *Figure 1B*. We compared the size of identity aftereffects for pairs of adapt and test faces that were taken from the different types of trajectories.

Adapt and test faces from same-race-average trajectories appear to be the same race, whereas adapt and test faces on the generic-average trajectories do not (see *Figure 2*). We, therefore, included the non-central trajectories (on which adapt-test pairs also appear to be the same race) to rule out the possibility that reduced aftereffects on the generic-average trajectories could somehow result from this mismatch in perceived race of the adapt and test faces. The non-central face identities were chosen to have the same perceptual distance to the target faces as the same-race average (of the same race) had to the target faces (see supplementary materials for details).

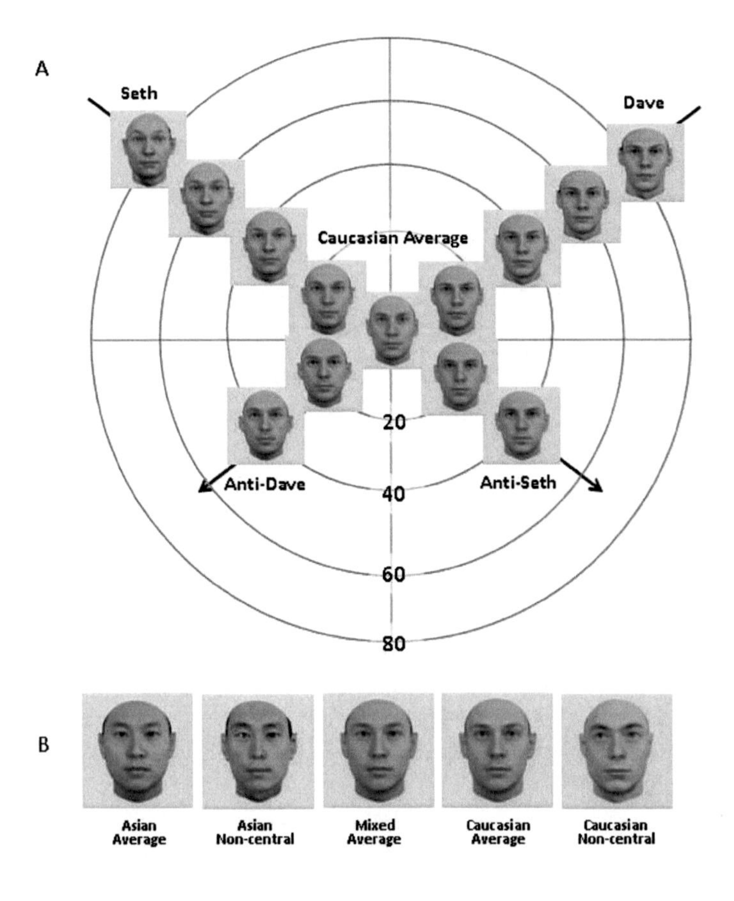

Figure 1. (A) Two identity trajectories through the same-race (here: Caucasian) average face, in a simplified two-dimensional face space. Antifaces were made by morphing the target identity towards and beyond the average face. On match trials, participants adapted to matching antifaces (e.g., adapt Anti-Dave, test Dave). On mismatch trials they adapted to non-matching antifaces from another identity (e.g., adapt Anti-Dave, test Seth). Numbers indicate percent identity strength (morph level). (B) The three average faces and the two "non-central" face identities used to create morph trajectories. The "non-central" and same-race averages for each race were matched on perceived similarity to the target faces.

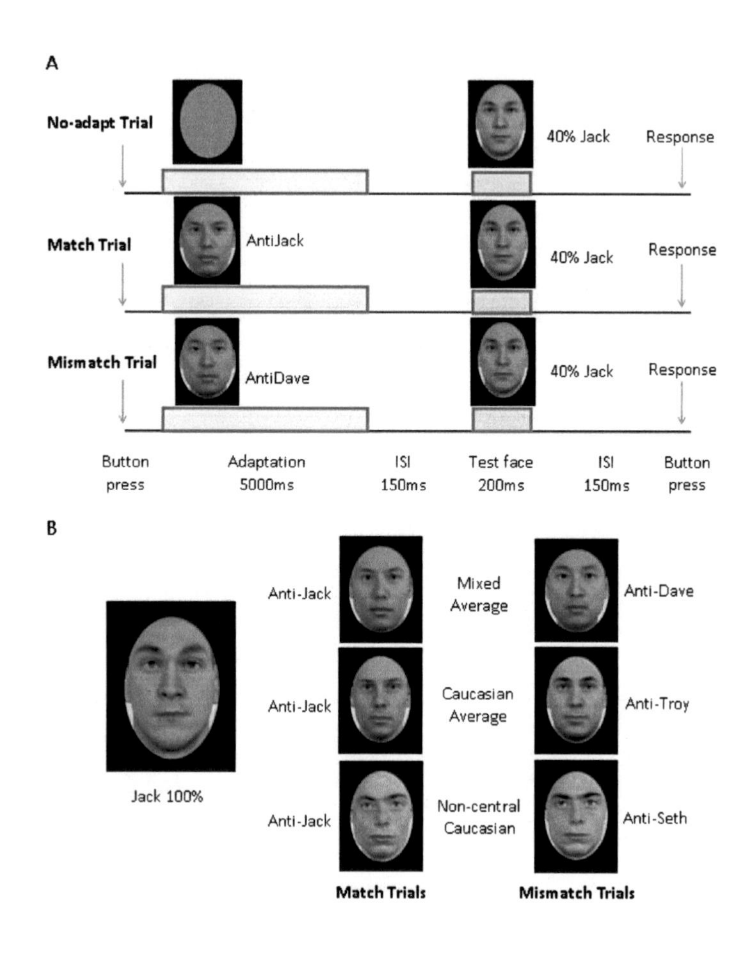

Figure 2. (A) Trial structure and conditions. Targets and antifaces shown come from a trajectory through the mixed-race average. (B) Example of matching and mismatching antifaces for each of the three trajectories for target face "Jack".

Since faces of any category might generally lie further from a generic face average than from a category specific one, test trajectories will likely be longer between target faces and generic averages than between target faces and category-specific averages. In this case, a larger aftereffect observed on a same-race trajectory might represent a smaller shift in perceptual distance in face space than a numerically smaller aftereffect on a (possibly longer) mixed-race trajectory. Additionally, when adapting and test stimulus become more similar, aftereffects can decrease as a function of decreasing perceptual contrast (Clifford, 2002; Robbins et al., 2007). We adjusted the size of the aftereffects to reflect any differences in test trajectory length and/or adapt-test similarity, as estimated from similarity ratings for antifaces and reference faces and for target faces and reference faces on each trajectory, respectively (see supplementary materials).

Aftereffects were measured, on each trajectory, as the difference between "match trials", where an antiface of the same identity as the test face was shown, and "mismatch trials", where an antiface of another identity was shown (cf., Pellicano, Jeffery, Burr & Rhodes, 2007; Rhodes et al., 2011). Larger aftereffects for adapt-test pairs made using the race-specific average than for those made using the generic average would indicate that Asian and Caucasian faces are coded using race-specific norms. If a single generic race-norm is used to code face identity, on the other hand, larger aftereffects for the mixed-race adapt-test pairs should be found. Aftereffects on the non-central trajectory should be smaller than on both other trajectories.

Another goal of the study was to test whether discrimination is enhanced around the norm. This was done by measuring identification thresholds in the absence of adaptation for target faces from the three different trajectories. If the trajectory that yields the largest aftereffects also yields the best identification performance, then we would conclude that discrimination is enhanced around the "true" norm. Many have proposed that face discrimination should be enhanced around the norm face, be it the "natural" one or the shifted norm after adaptation (e.g., Wilson, Loffler, & Wilkinson, 2002; Rhodes, Maloney, Turner, & Ewing, 2007; Rhodes, Watson, Jeffery, & Clifford, 2010), as sometimes found in lower-level vision (Kohn 2007). However, the evidence is equivocal so far, with several studies finding no evidence that adaptation to either the average face (i.e., the "natural" norm) or another face identity (to shift the natural norm) improved performance around the old or new norm (Jaquet, Rhodes, & Clifford, 2005; Ng, Boynton, & Fine, 2008; Rhodes, Maloney et al., 2007). In contrast, others have found enhanced discrimination around an adapted state. For example, discrimination of synthetic radial frequency faces was better around the average than a non-central location of their face space (Wilson et al., 2002). Adaptation to the average of a natural-looking face population enhanced identification of faces from the adapted, relative to an unadapted population (in this case Asian and Caucasian faces) (Rhodes, Watson, et al., 2010). Adaptation to a male or female face selectively enhanced gender discrimination for faces from the respective category (at least for faces from the same identity continuum, see Yang, Shen, Chen, & Fang, 2010). And finally, adaptation can enhance discrimination of face views around an adapted viewpoint (Chen, Yang, Wang, & Fang, 2010). In the current study, we asked whether participants are better at discriminating between realistic face identities that vary around the true psychological norm rather than a non-central location in face-space. Better discrimination around the norm of a face category would be direct evidence of a functional benefit of norm-based coding in face perception.

Methods

Participants

Twenty-nine Caucasian adults (eighteen female, eleven male) between 18 und 35 years of age were recruited from the University of Western Australia and received either credit points or cash reimbursement for their participation.

Stimuli and Apparatus

All faces were derived from scanned 3D-heads from the Max Planck face database in Tübingen (http://faces.kyb.tuebingen.mpg.de/index.php). The "Morphable Model" algorithm developed by Blanz and Vetter (Blanz, 2000; Blanz & Vetter, 1999) was used to generate stimuli. Four Asian and four Caucasian male faces were chosen as target faces, based on easy discriminability, by the experimenters.

Asian and Caucasian averages were created by morphing together 20 Asian and 20 Caucasian male faces (including the four target faces), respectively. A generic average was created from all 40 faces of both races. The non-central reference face was another face identity of the same race as the target faces (i.e., Asian for the Asian target faces and Caucasian for the Caucasian targets). The two non-central faces were chosen on the basis of similarity ratings, to match the perceptual distance between each non-central face and the target faces of the corresponding race to the perceptual distance between the same-race averages and the targets (see supplementary materials for details).

We created three different antifaces and thus three different trajectories for each target face by morphing their texture and shape toward and beyond each of the three reference faces *(see Figure 1)*: (1) A same-race trajectory, where the target faces were morphed through their same-race average (i.e., Asian for Asian target faces and Caucasian for Caucasian targets); (2) a mixed-race trajectory, where the average made from both Asian and Caucasian faces was used; and (3) a non-central trajectory, using a non-central same-race face instead of an average face. Morphing was done from each target face (100% identity strength) through one of the reference faces (0%) to a -50% antiface (no further, since stronger antifaces showed distortions). Each test trajectory consisted of seven versions of the target face, varying in identity strength: -20%, 0%, 20%, 40%, 60%, 80%, 100% (negative values represent antifaces).

All face stimuli were devoid of secondary cues such as hair and make-up. They were presented in color, surrounded by a black oval mask that hid the inner hair line and part of the ears (see *Figure 2*) but not the face outline, on a black background. A grey oval of the same size and overall luminance as the antifaces was used on no-adaptation trials. The adapt faces measured approximately 125% of the size (8.6° x 7.0°) of the test faces (7.2° x 5.2°), at a viewing distance of about 57 cm.

Experiments were run on a 20-inch LCD screen (1680 by 1050 pixels resolution) iMac OS X, version 10.5.6, using SuperLab 4.0.6 software. Participants always responded using a standard computer keyboard.

Similarity Ratings

Similarity ratings were obtained from additional participants to assess (and adjust aftereffects for) any differences in (a) perceptual length, and (b) "extremity" of the adaptor faces relative to the average faces for the three kinds of trajectories. Additional participants rated the similarity of target - average pairs (80% vs. 0%) and the similarity of antiface – average (-50% vs. 0%) pairs on each trajectory. Face pairs were presented sequentially on the computer screen, and participants answered using a 7-point scale with 1 labeled "not at all similar" and 7 labeled "very similar". See supplementary material for details of these rating experiments.

Procedure

Each session began with a training phase, in which participants learned four target face identities, either four Asian male faces or four Caucasian male faces. In the subsequent identification phase, participants had to identify the same four target faces at various identity strengths, after adapting to either a matching or mismatching antiface. Trials for the three trajectories were intermixed. Identification was also tested in the absence of adaptation. Adaptation and no-adaptation trials were randomly inter-mixed and the whole identification phase was repeated three times, in separate sessions, each lasting 45-60 minutes. Target race was the only between-participants factor; all other factors were varied within participants.

Training. At the beginning of each session, participants were given a print-out of the four target faces (at 100% identity strength) and their corresponding names, and were asked to assign characteristic adjectives (e.g., likable, trustworthy, arrogant) from a list to each face identity. They were told to take as much time as necessary and when they felt confident they could tell the faces apart and identify each by name, they moved on to practice identifying the faces on the computer. This practice consisted of three blocks of trials in which the same four target faces (also at 100% identity strength) were presented for an unlimited amount of time (in the first block), 500ms (second block) and 200ms (third block). Participants answered using labeled keyboard keys and received visual accuracy feedback at the end of each trial. Each face was shown four times in random order in each practice block (i.e. 16 trials per block), and blocks were repeated if necessary until participants could correctly identify all faces with 100% accuracy. The print-out of the faces was kept in view during the first block, but participants were advised that it would be removed during the experiment so they should feel certain that they could identify the faces without it. Finally, in two additional blocks of trials, participants practiced identifying the four test faces at weaker identity strengths (80% and 60%), taken from all three trajectories (same-race, mixed-race, non-central). Exposure duration was unlimited in the first block and 200ms in the second block. Each face was shown once at each identity strength for each trajectory, i.e., 4 identities x 3 trajectories x 2 identity strength levels = 24 trials per block.

Identification. For each trajectory (same-race, mixed-race, non-central), there were two adapting conditions (*see Figure 2A*): (a) match adapt, in which the identity of the antiface used as the adapting stimulus corresponded to the target face, and (b) mismatch adapt, in which a randomly assigned antiface of one of the other three targets faces was used as the adapting stimulus. The non-matching antifaces were pre-assigned to each of the three identities across trajectories (i.e., anti-Dan was used as mismatching adaptor for Jack in one trajectory, for Seth in another and for Troy in the third one), and held constant within each trajectory *(see Figure*

2B). On no-adaptation identification trials, a grey oval stimulus replaced the adapting antifaces, to maintain the trial sequence. It was the same average size and average luminance as the antifaces (see *Figure 2A* for trials of all conditions). For each testing session, there were 216 trials, consisting of 4 target identities x 3 testing conditions (match, mismatch, no-adapt) x 6 identity strengths (-20%, 0%, 20%, 40%, 60%, 80%) x 3 trajectories (same-race, mixed-race, non-central). Trials were presented in random order. Each trial consisted of 5000 ms exposure to an antiface adaptor or a grey oval, followed by a 150 ms ISI, followed by a test face for 200 ms, followed by a 150 ms blank ISI, followed by a response screen asking the participant to indicate the identity of the target. Participants initiated each trial by pressing the spacebar when they were ready and used the same labeled keyboard keys to answer as during training. Participants were told to be as accurate as possible even though some of the faces would be difficult to identify, and that they should make their best guess when uncertain. There were self-timed rest breaks after every 36 trials.

Results

Participants' identification responses were scored correct if they corresponded to the identity from which the test face was made. Following Leopold et al. (2001), a correct response was arbitrarily assigned to each presentation of a 0% identity strength face, with each of the four target identities assigned equally often, to measure "performance" on these trials. The data of two (female) participants was removed from the analysis; one because her recognition performance was less than three standard deviations below the group mean in six out of nine conditions (no other participant was that far from the group mean in any condition), the other one because she did not complete all three experimental sessions.

Aftereffects

For each participant, the mean proportion correct was calculated and plotted as a function of identity strength for each trajectory and adaptation condition (match, mismatch). Cumulative Gaussians were fitted to each identification curve using GraphPad Prism version 5.00 for Windows (GraphPad Software, San Diego California USA, www.graphpad.com). The fits were good (mean $R^2 = 0.931$, SD = 0.062, range = 0.668 to 1.000, N = 27 participants x 3 trajectories x 3 adapt conditions). Mean identification curves for match and mismatch trials (averaged across participants) for all 3 trajectories (same, mixed, non-central) are shown in *Figure 3.* On all trajectories, performance for match trials was better than for mismatch trials.

Figure 3. Mean identification performance as a function of identity strength after adapting to matching and mismatching antifaces for each trajectory (same-race, mixed, non-central). Fitted cumulative Gaussians are shown.

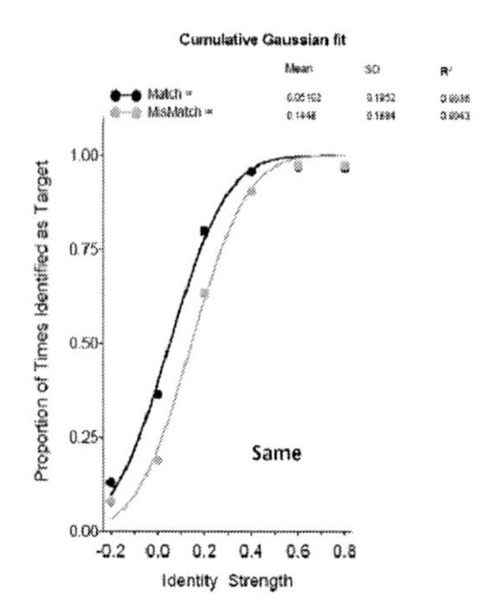

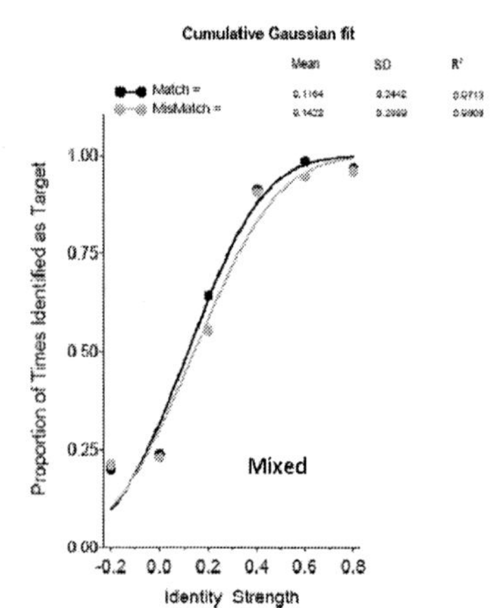

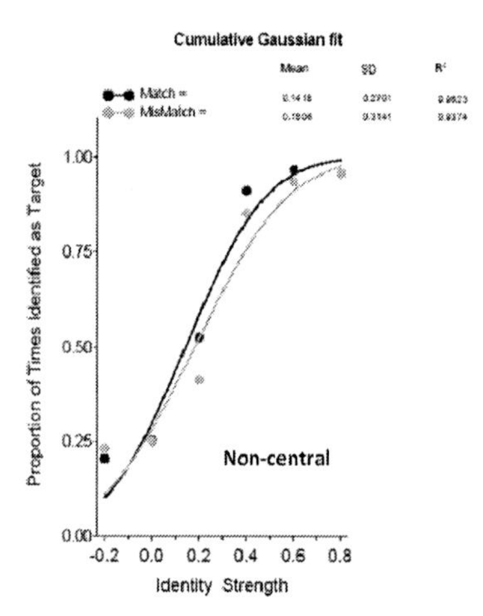

For each participant, we calculated an adaptation aftereffect for each trajectory by subtracting identification thresholds on match trials from identification thresholds on mismatch trials. We used the means of the fitted Gaussians as the identification thresholds (see *Figure 4A and Table 1*). The aftereffect *(Figure 4B)* was bigger for the same trajectory than for the other two trajectories, while the aftereffects for the mixed and non-central trajectories did not differ. This pattern was confirmed by a repeated measures Analysis of Variance (ANOVA), with target race (Asian, Caucasian) as a between-subjects factor and trajectory (same-race, mixed, non-central) as a within-subject factor. A main effect of trajectory, $F(2,50) = 7.480$, $p = 0.001$, *partial* $\eta^2 = 0.230$, indicated that the size of the aftereffect differed between the three conditions. There was no significant main effect of target race, $F(1,25) = 1.633$ $p = 0.213$, *partial* $\eta^2 = 0.061$, and no interaction, $F(2,50) = 0.695$, $p = 0.504$, *partial* $\eta^2 = 0.027$. Planned pair-wise t-tests confirmed that the aftereffect was significantly larger on the same-race trajectory than on both other trajectories [*same vs. mixed:* $t(26) = 3.307$, $p = 0.003$; *same vs. non-central:* $t(26) = 3.592$, $p = 0.001$], suggesting that the antifaces on the same-race trajectories are the ones that lie truly opposite to the targets in face space. The aftereffects for the mixed and non-central trajectories were not significantly different from each other [$t(26) = 0.268$, $p = 0.791$].

Table 1

Aftereffects calculated from means of cumulative Gaussians as a function of face race and trajectory.

Race of targets	Trajectory	Mean	SE
Caucasian	Same	0.082	0.016
	Mixed	0.028	0.018
	Other	0.044	0.024
Asian	Same	0.121	0.018
	Mixed	0.045	0.016
	Other	0.039	0.016

A: Identification thresholds

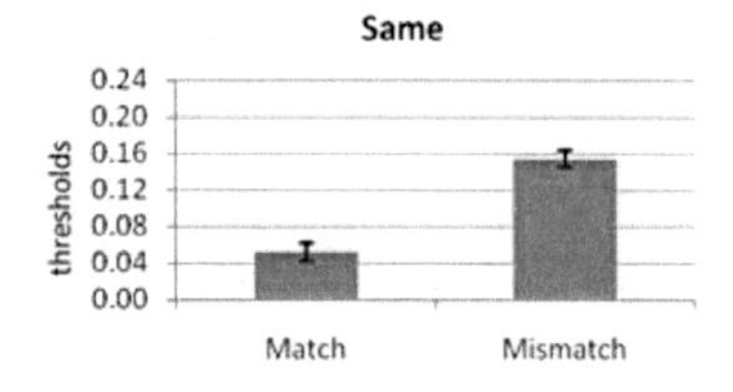

B: Aftereffect measured from thresholds

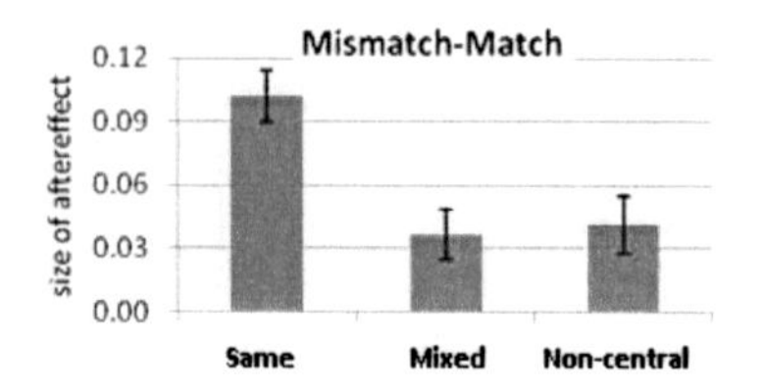

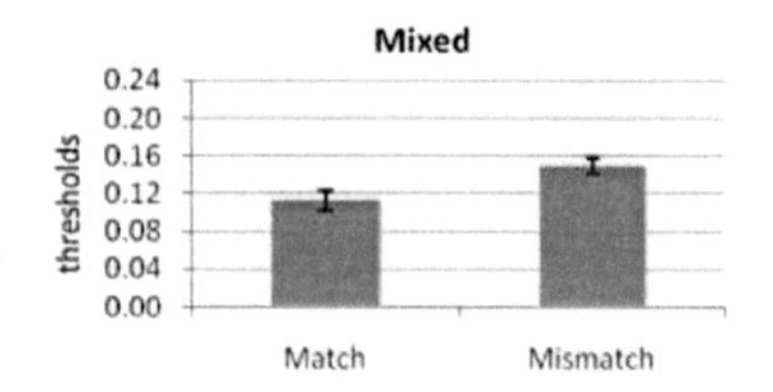

Figure 4: (A). Mean identification thresholds (means of fitted Gaussians) after adapting to matching and mismatching antifaces for each trajectory. (B) Size of aftereffects for each trajectory.

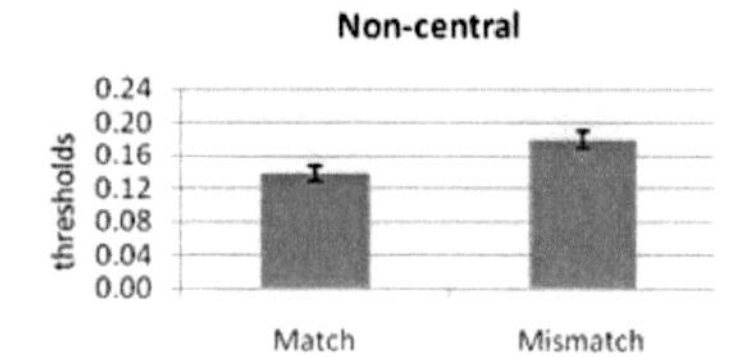

Correction for differences in trajectory length. Mean similarity ratings for target - average pairs (80% vs. 0%) (see supplementary materials) were 30% lower for mixed (M = 2.52, SE = 0.24) and 3% lower for non-central (M = 3.52, SE = 0.22) trajectories, compared to same-race trajectories (M = 3.61, SE = 0.21), indicating that the mixed and non-central test trajectories were longer than the race-specific ones. We therefore scaled the mixed (multiplying by 1.30) and non-central (multiplying by 1.03) trajectory aftereffects and repeated the two-way ANOVA. Aftereffects remained larger for same-race (M = 0.102, SE = 0.012) than for mixed (M = 0.048, SE = 0.015) and non-central (M = 0.043, SE = 0.014) trajectories (*Table 2*). There was a significant effect of trajectory, $F(2,50) = 4.844, p = 0.012$, *partial* $\eta^2 = 0.162$, but no main effect of face race, $F(1,25) = 1.678, p =$

0.207, partial η^2 *= 0.063,* and no interaction, *F(2,50) = 0.570, p = 0.569, partial* η^2 *= 0.022.* Pairwise t-tests showed that aftereffects remained significantly larger for same-race than for mixed and non-central trajectories *[same vs. mixed: t(26) = 2.385, p = 0.003; same vs. non-central: t(26) = 3.457, p = 0.002]*, with no significant difference between mixed and non-central trajectories *[t(26) = 0.219, p = 0.829]*.

Table 2

Aftereffects from Table 1, adjusted for different length of trajectories. See suppl. materials for details.

Race of targets	Trajectory	Mean	SE
Caucasian	Same	0.082	0.016
	Mixed	0.036	0.024
	Other	0.046	0.025
Asian	Same	0.121	0.018
	Mixed	0.059	0.021
	Other	0.041	0.016

Correction for differences in adaptor-test contrast. Mean similarity ratings for antifaces (-50%) on same-race (M = 5.33, SE = 0.20) and mixed trajectories (M = 5.13, SE = 0.17) were comparable, but antifaces on non-central trajectories (M = 6.21, SE = 0.17) were significantly more similar to their reference faces (see supplementary materials). The smaller aftereffect on the non-central trajectory could therefore potentially be attributed to the use of less extreme adaptors than on the other trajectories. However, when we re-scaled the previously scaled aftereffects (see above) again, according to these ratings, and repeated the ANOVA, the results remained exactly the same (see supplementary materials for more details and *Table 3* for means and SEs). There was a significant main effect of trajectory, *F(2,50) = 4.323, p = 0.019, partial* η^2 *= 0.147*, but no main effect of face race, *F(1,25) = 1.443, p = 0.241, partial* η^2 *= 0.055,* and no interaction , *F(2,50) = 0.557, p = 0.577, partial* η^2 *= 0.022.* Pairwise t-tests showed that aftereffects remained larger for same-race than for mixed and non-central trajectories *[same vs.*

mixed: t(26) = 2.523, p = 0.018; same vs. non-central: t(26) = 2.919, p = 0.007], with no significant difference between aftereffects for mixed and non-central trajectories *[t(26) = 0.110, p = 0.913].*

Table 3

Aftereffects from Table 1, adjusted for different length of trajectories and extremity of adaptor face. See suppl. materials for details.

Race of targets	Trajectory	Mean	SE
Caucasian	Same	0.085	0.016
	Mixed	0.036	0.024
	Other	0.053	0.029
Asian	Same	0.126	0.019
	Mixed	0.059	0.021
	Other	0.048	0.019

Identification Performance. To assess whether identification performance was best around the "true norm", we plotted performance as a function of identity strength for each trajectory, for each participant. Mean identification curves for the three trajectories are plotted in *Figure 5A.* Overall identification accuracy was highest (i.e., identification thresholds were smallest) on the same-race trajectory, followed by the mixed trajectory, and poorest on the non-central trajectory *(also see Table 4).* The same two-way ANOVA as above showed a significant effect of trajectory, $F(2,50) = 27.579, p < 0.001$, *partial* $\eta^2 = 0.525$, but no main effect of target race, $F(1,25) = 2.361, p > 0.135$, *partial* $\eta^2 = 0.086$, and no interaction, $F(2,50) = 3.633, p > 0.065$, *partial* $\eta^2 = 0.127$. Pair-wise t-tests showed that all three pairwise differences were significant *[same vs. mixed: t(26) = 2.79, p = 0.010; same vs. non-central: t(26) = 6.75, p < 0.001; mixed vs. non-central: t(26) = 4.08, p < 0.001].*

Correction for differences in trajectory length. When we adjusted identification thresholds in the same way as aftereffects, to correct for differences in length of the three trajectories (see above), the pattern remained the same: Thresholds were still smallest for the same-race trajectory, followed by the mixed trajectory and non-central trajectory (see *Table 5*). The ANOVA showed a significant effect of trajectory, $F(2,50) = 27.414, p < 0.001, partial\ \eta^2 = 0.523$, but no main effect of target race, $F(1,25) = 1.971, p > 0.170, partial\ \eta^2 = 0.073$, and no interaction, $F(2,50) = 3.158, p > 0.085, partial\ \eta^2 = 0.112$. Pair-wise t-tests showed that differences between the same-race and both other trajectories were significant [*same vs. mixed:* $t(26) = -5.77, p < 0.001$; *same vs. non-central:* $t(26) = 7.07, p < 0.001$], but not between the mixed and the non-central trajectories [$t(26) = 1.185, p = 0.247$].

Table 4

No-adapt identification thresholds as a function of face race and trajectory.

Race of targets	Trajectory	Mean	SE
Caucasian	Same	0.132	0.016
	Mixed	0.164	0.011
	Other	0.263	0.021
Asian	Same	0.131	0.008
	Mixed	0.168	0.016
	Other	0.200	0.012

Table 5

No-adapt identification thresholds from Table 4, adjusted for differences in trajectory length.

Race of targets	Trajectory	Mean	SE
Caucasian	Same	0.132	0.016
	Mixed	0.214	0.014
	Other	0.271	0.216
Asian	Same	0.131	0.008
	Mixed	0.219	0.021
	Other	0.206	0.013

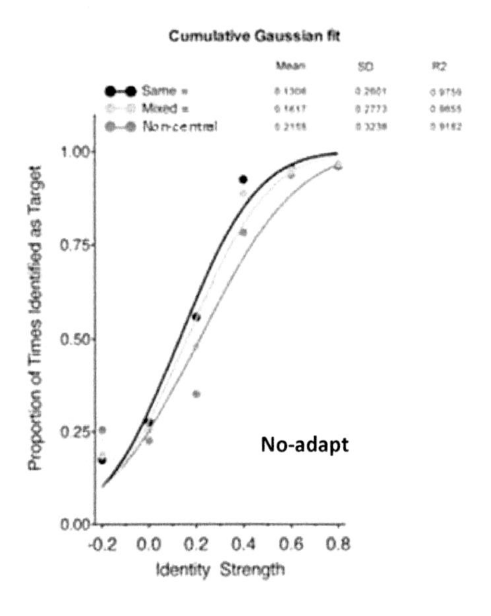

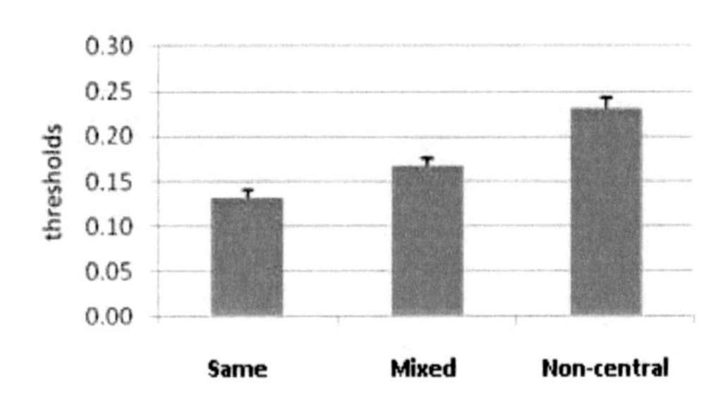

Figure 5. (A). Mean identification performance as a function of identity strength for each trajectory (same-race, mixed, non-central). Fitted cumulative Gaussians are shown (B). Mean identification thresholds (means of fitted Gaussians) for each trajectory (same-race, mixed, non-central).

Discussion

Judging the race of an individual in every-day life of is not merely about classifying faces into visually distinct categories based on physical facial differences. Faces of different races also represent meaningful biological and social categories, and belonging to one or the other might have important social implications. In this study, we therefore investigated how faces of different races are represented in the brain, that is, whether they are coded relative to race-specific norms or relative to a common, generic face norm. We found larger face identity aftereffects for adapt-test pairs that lie opposite a race-specific average than opposite either a generic mixed-race average or another identity from a non-central location in face-space. Since previous studies have shown that aftereffects are largest when adapt-test pairs lie opposite the average in face-space (Leopold et al., 2001; Rhodes & Jeffery, 2006), our results strongly suggest that the race of a face plays a role for encoding its identity, and that faces of different races are coded using race-specific norms.

Race-contingent figural face aftereffects, where adaptation to oppositely distorted Chinese and Caucasian faces induces opposite changes in the perceived normality of Chinese and Caucasian test faces suggest that different norms are maintained for faces of different races (Jaquet et al., 2007; 2008). The present results explicitly implicate race-specific norms in coding the identity of faces. A similar adaptation paradigm has also been used to show that identity is coded relative to norms selective for the sex of each face (Rhodes et al., 2010). Taken together, these findings suggest that face coding mechanisms are tuned to capture variation within visually distinct and socially important categories (see also Little et al., 2008).

The question remains how category-specific norms are implemented in face space. Since faces of different races and sexes share many visual properties, a functional and neural architecture in which these categories are represented in distinct spaces and coded by completely distinct neural populations is implausible. Instead, for male and female faces, a "dissociable coding model" has been proposed, in which faces are represented in a single face space containing common dimensions that are represented in all faces, as well as category-selective dimensions that are most useful for differentiating faces of one particular category (see Rhodes & Jaquet, 2010, Rhodes et al., 2010). We propose a similar system for faces of different races, with some common dimensions that are shared by faces of all races, and some race-selective dimensions.

It is possible that our mixed-race average face might not represent the true generic norm for face race in general. This average was created by morphing together only Asian and Caucasian faces which may not represent a realistic generic race norm, since there are more than two race categories. It is also possible that even our Caucasian/Asian average lies somewhere between Asian and Caucasian faces but not exactly in the middle as we assumed by morphing an equal number of faces of each race. Nevertheless, an average face made from two races morphed together should still be closer to the central tendency of all race categories (in the absence of any sub-categorization) than is a Caucasian or an Asian average alone. Therefore, if a generic norm is used as a perceptual reference, we should still have found a bigger aftereffect for the mixed-race trajectory than for race-specific ones. But instead we found the reverse effect, providing evidence that faces of different races form distinct (although overlapping) perceptual categories and that these categories are coded using different norms.

Could it be problematic that our race-specific averages consisted of only twenty faces, while the generic average face was created by morphing together all forty faces? We suggest not, because increasing the number of faces included in the generic average should produce a face that is even closer to the central tendency of all faces, making the mixed-race average even "more average". Moreover, given that a generic norm that is built up over experience will naturally be based on more faces than category-specific norms of any kind, the difference in the number of faces used to create our averages in fact reflects the reality of such norm faces. Thus, again, if faces were not categorized into Asian and Caucasian faces, the mixed average should be closer to the center of the face-space and we should have found larger aftereffects for mixed-race trajectories. However, we found larger aftereffects for same-race trajectories, suggesting that the mixed-race average is not the true psychological norm.

To more precisely control for differences in perceptual distances on the three trajectories, we obtained similarity ratings between averages and target faces on all trajectories and adjusted the aftereffects accordingly. To control for differences in adaptor-test similarity and thereby for perceptual contrast between the two, which could lead to differences in the size of the aftereffect (Clifford, 2002; Robbins et al, 2007), we also adjusted the aftereffects based on similarity ratings of the perceptual distance between reference faces and antifaces on each trajectory. Aftereffects remained larger on race-specific trajectories after correction for differences in trajectory length and in extremity of the adaptor face.

Our results also provide direct evidence of a functional advantage of norms in face perception. Identification was better around the true perceptual (i.e., race-specific) norm than around a mixed-race average or another non-central face identity (see *Figure 5*), and the advantage remained after correcting for differences in trajectory length. This result replicates Wilson et al.'s (2002) earlier finding using synthetic radial frequency faces. It suggests that norm-based coding of face identity may facilitate successful recognition, by highlighting what is distinctive about each individual face and minimizing the processing of redundant facial information. The use of distinct (although overlapping) norms for visually distinct face categories, as indicated by our results, might be even more economical, since neural responses increase with distance from the norm, and faces will generally lie further from a generic than a category-specific average.

Another advantage of category-specific norms might be that they can be selectively updated so that changes in visual experience with one face population can selectively re-tune coding of faces within that population. Experience can also shift category boundaries closer to the more familiar category (Webster, Kaping, Mizokami, Duhamel, 2004), which might serve to increase discriminability within that category. One might ask then why, since all our participants were Caucasian, we did not find better performance for Caucasian than Asian face identities. We suggest that it is because our participants extensively learned and were tested on only four individual faces of each race, so that they were highly familiar with the targets. When only a few highly familiar faces must be discriminated, other-race effects are not generally observed (Jaquet, Rhodes & Hayward, 2008; Rhodes, Watson, Jaquet, Winkler, & Clifford, 2004).

It remains to be tested how norms for faces of distinct race categories differ between races and between individuals. Since depending on the environment one grows up in, we all have very different experience with only one or sometimes several races, everybody's long-term race norms might also develop very specifically. Correlating personal experience with different races and performance in high-level face adaptation experiments might shed some light on the question how norms are derived over time, and how this process influences perception of and identification performance for these visually and socially different groups of faces.

In summary, our data add to a growing literature showing that exposure to faces biases subsequent perception of novel faces (e.g., Leopold et al 2001; Rhodes et al., 2003, 2004; Webster et al., 2004), presumably by shifting a norm that is used to encode faces in the brain. Faces as a category can be divided into subcategories based on appearance, and here we show that faces of different races are coded against their respective category-specific norm face, rather than a generic single norm. The face space model may thus be best conceptualized as a space containing category-selective, as well as generic, face dimensions. In addition, our findings suggest a functional role for long-term face norms abstracted from adaptation to populations of faces over long time periods.

Acknowledgements

This research was funded by the Australian Research Council and a project grant from the German Academic Exchange Service awarded to Regine Armann, and supported by the Max Planck Society. We wish to thank Isabelle Bülthoff, Christian Wallraven and Johannes Schultz for help with stimulus collection and the face database.

Supplementary Materials: Similarity Ratings

(1) Ruling out category priming

Faces of different races represent distinct categories and might be coded in different locations in face space (e.g., Jaquet et al., 2008). This raises the question whether adaptation with an antiface of the same race as the target face (which is true for the same-race trajectory in our experiment) gives a recognition advantage over adaptation to an antiface of a different race (as on the mixed-race trajectory). In this case, better performance on the same-race trajectory that we found in this study might just be due to priming of the correct face category on this but not on the mixed-race trajectory. To rule out such an account, we introduced the non-central trajectory, where the antiface is of the same race as the target faces (as on the same-race trajectory), but is created relative to an individual identity that is taken from a non-central location in face space. To make sure that the similarity between the targets and the reference face are the same on both trajectories (and that aftereffects on these trajectories are thus comparable), we chose these non-central reference faces on the basis of similarity ratings. The perceptual distance between the target faces and the same-race average was thus matched to the perceptual distance between the targets and the two non-central reference faces that were used in the main study.

Method

Participants

Fourteen young Caucasian adults from the student pool of the University of Western Australia volunteered to do the rating experiment. Eleven participants rated Caucasian and nine participants Asian faces (with six students rating only Caucasian target faces, four only Asian target faces, and five students doing both versions of the experiment, with a time lag of at least 6 days between sessions). None participated in the main experiment.

Stimuli

Thirteen Asian and thirteen Caucasian faces from the pool that was used to create the average faces (twenty Asian and twenty Caucasian male faces) were used for this rating experiment.

Procedure

The task was run on a 20-inch LCD screen iMac OS X, version 10.5.6. Stimuli were presented using the presentation software Superlab 4.0.6, at a screen resolution of 1680 by 1050 pixels. Participants were first presented with instructions on the computer screen explaining what would be required of them.

Thirteen faces from each set of twenty faces used to create the Asian and Caucasian averages were paired with each of the four target faces of the same race set (Asian faces with Asian targets and Caucasians with Caucasians, respectively). All faces, including the target faces, were also paired with their own-race average. Participants were presented with one face pair at a time, along with a rating scale, ranging from 1 (not at all similar) to 7 (very similar). Every face pair (69 per race) was shown twice, once in each left-right arrangement. Participants were instructed to answer as quickly as possible and to use the full range of the scale, using the number keys on the computer keyboard. To expose them to the appearance and range of stimuli, participants were shown 22 other random single faces at the beginning of the experiment. They had no answer to give and these faces were not used again later for the rating. The whole experiment took about 10 minutes to complete.

Results

The perceptual similarity ratings for each face pair were averaged across participants, for each race separately. The average distance between the Asian targets and the Asian average face was 3.45 (SE = 0.26). The face whose ratings came closest to that had an average distance to the targets of 3.53 (SE = 0.36). The average distance between the Caucasian targets and the Caucasian average face was 3.92 (SE = 0.24). The face whose ratings came closest to that had an average distance to the targets of 3.66 (SE =0.40). These two faces with an average distance to the targets that was very close to the one between the targets and their own-race average face were thus used in lieu of an average face to create the 'non-central' morph trajectory in the adaptation experiment. Pair-wise t-tests show that the similarity ratings for targets and average and the ratings for targets and these two non-central faces did not differ significantly *[Asian faces: $t(8) = -0.25$, $p = 0.809$ (SE = 0.320); Caucasian faces: $t(10) = 1.09$, $p = 0.312$ (SEM = 0.244)]*.

(2) Length of trajectories and extremity of adaptors

In the main experiment, we found that aftereffects were larger for same-race than mixed-race trajectories. Since aftereffects are larger for opposite than non-opposite adapt-test pairs, we argued that this suggests that identity is coded using a same-sex average. However, larger aftereffects on same-race trajectories could also be explained by two alternative accounts. First: The target faces are likely to be more similar to an average face of the same race than to a mixed-race average. If so, the larger aftereffects that we found for (shorter) same-race trajectories might actually represent smaller absolute shifts in perceptual distance than the smaller aftereffects observed for (possibly longer) generic trajectories.

Second: Since the antifaces are also likely to be more similar to a same-race than a mixed-race average, adaptors might be less extreme on same-race than mixed-race trajectories. While face aftereffects are generally larger for more extreme adaptors (Robbins et al., 2007), we found the opposite pattern: larger aftereffects on same-race trajectories. Hence, the expected difference in adaptor extremity would be in the wrong direction to explain our results.

To assess the length of the three trajectories and to estimate adaptor-test similarity for each trajectory, we obtained ratings of similarity of the target faces to the averages (80% vs 0% pairs) and of the adapting antifaces to the averages (-50% vs 0% pairs).

Method

Participants

Twelve young Caucasian adults were recruited from the student pool at the University of Western Australia. None had participated in the main experiment.

Stimuli

The same face stimuli as in the main experiment were used. For the present rating study, stimuli consisted of the four target faces (for each race, respectively) morphed with each of the three reference faces, at an identity strength of 80%. For each of the resulting faces (12 faces in each race), the antiface (-50%) was included, as well as the three reference faces for each race. There were thus 27 faces for each race.

Procedure

The task was run on a 20-inch LCD screen iMac OS X, version 10.5.6. Stimuli were presented using the presentation software Superlab 4.0.6, at a screen resolution of 1680 by 1050 pixels. Participants were first presented with instructions on the computer screen explaining what would be required of them.

For each trajectory, two comparisons were made: target face (80%) vs. reference (0%), and reference (0%) vs. antiface (-50%). Pairs of faces were shown sequentially, and each comparison was presented twice with sequence order presentation alternated. In total, there were 96 trials (3 trajectories x 4 target faces x 2 races, x 2 comparisons, x 2 sequence orders), presented in two blocks of 48 trials of only one race, but with trajectories presented intermixed (randomized). Block presentation order was counterbalanced across participants.

Each block consisted of an initial viewing phase followed by a rating phase. In the viewing phase, participants were shown all pair combinations and were instructed to look closely at each face pair to get a general idea of how the pairs differ in similarity. Once all pair combinations were viewed, participants were instructed on the screen to look at all pairs again and this time rate them for similarity, using a numbered 7-point scale ranging from 1 (not at all similar) to 7 (very similar). Participants were instructed to answer as quickly as possible and to use the full range of the scale, using the number keys on the computer keyboard. The whole experiment (i.e., both blocks of Asian and Caucasian faces, respectively) took approximately 15 minutes to complete.

Results

Similarity target – reference faces (80% vs 0%)

Mean similarity ratings were calculated for each trajectory and face-race for each participant. A two-way ANOVA with trajectory as within-subjects factor and face race as between-subjects factor was conducted on the mean similarity ratings. There was a significant effect of trajectory *[$F(2,44) = 14.500, p < 0.001$, partial $\eta^2 = 0.397$]*. Pair-wise t-tests show that pairs from same-race trajectories (Mean = 3.61, SE = 0.21) and pairs from non-central trajectories (Mean = 3.52, SE = 0.22) were rated as significantly more similar than pairs from mixed-race trajectories (Mean = 2.52, SE = 0.24; *same vs. mixed: $t(23) = 3.860, p = 0.001$; non-central vs. mixed: $t(23) = 0.4220, p < 0.001$*). Pairs from non-central trajectories, however, had the same perceptual similarity as pairs from same-race trajectories *($t(23) = 0.759, p = 0.456$)*. This is in accordance with the fact that the non-central reference faces were explicitly chosen on the basis of similarity ratings, to match their distance to the targets to the distance between the targets and the same-race averages. There was no significant effect of target race *[$F(1,22) = 0.108, p = 0.746$, partial $\eta^2 = 0.005$]* nor an interaction with trajectory *[$F(2,44) = 0.936, p = 0.4$, partial $\eta^2 = 0.041$]*. See Table 1 for means and SEs.

Table 1. Mean (SE) similarity of **target faces** to same-race and mixed-race **reference faces** (80% vs. 0%)

	Reference / Trajectory		
Face Race	same-race	mixed-race	non-central
Asian	3.73 (0.36)	2.40 (0.25)	3.34 (0.38)
Caucasian	3.50 (0.23)	2.64 (0.41)	3.70 (0.25)
All	3.61 (0.21)	2.52 (0.24)	3.52 (0.22)

Correction for differences in test trajectory length

The similarity ratings for the target-reference (80% vs. 0%) pairs were used to estimate and adjust any difference in test trajectory lengths. Target faces were rated as 30% less similar to the generic than the same-race averages (2.52 vs. 3.61) and 3% less similar to the non-central than the same-race average (3.52 vs. 3.61). The generic-trajectory aftereffect was therefore scaled (multiplying) by 1.3, and the non-central-trajectory aftereffect was multiplied by 1.03. Then, the effect of trajectory was reexamined in a two-way ANOVA with trajectory as repeated measures factor and face race as between-subjects factor. The ANOVA revealed a significant effect of trajectory *[$F(2,50) = 4.844$, $p = 0.012$, partial $\eta^2 = 0.162$]*, but no main effect of face race *[$F(1,25) = 1.678$, $p = 0.207$, partial $\eta^2 = 0.063$]* and no interaction *[$F(2,50) = 0.570$, $p = 0.569$, partial $\eta^2 = 0.022$]*. Pair-wise t-tests show that aftereffects remained larger for same than for mixed and non-central trajectories *[same vs. mixed: $t(26) = 2.385$, $p = 0.003$; same vs. non-central: $t(26) = 3.457$, $p = 0.002$]*, whereas the aftereffects for mixed and non-central trajectories were not significantly different from each other *[$t(26) = 0.219$, $p = 0.829$]*.

<u>Correction for differences in extremity of adaptors</u>

Antifaces were rated as more similar to non-central (M = 6.21, SE = 0.17) than to mixed (M = 5.13, SE = 0.17) and race-specific averages (M = 5.33, SE = 0.20). Pair-wise t-tests showed that the difference in similarity between non-central trajectories and both other trajectories was significant *[same vs. non-central: $t(23) = -4.892$, $p < 0.001$; mixed vs. non-central: $t(23) = 4.308$, $p < 0.001$]*, but not the difference on same-race and mixed-race trajectories *[$t(23) = 0.866$, $p = 0.396$]*. There was a significant main effect of trajectory *[$F(2,44) = 12.697$, $p < 0.001$, partial $\eta^2 = 0.366$]*, but no main effect of face race *[$F(1,22) = 0.554$, $p = 0.465$, partial $\eta^2 = 0.025$]* and no interaction *[$F(2,44) = 0.459$, $p = 0.635$, partial $\eta^2 = 0.020$]*.

Since the non-central antifaces are more similar to the reference faces, they are less extreme adaptors than the antifaces on the other trajectories. Smaller aftereffects for non-central antifaces could in principal be explained by these differences. We thus scaled the aftereffects again, relative to these differences. Here, contrary to the trajectory length measured above, a higher value (i.e., more similar antiface – reference pairs) is expected to lead to a weaker aftereffect. The race-specific antiface is 4% more similar to the average than the mixed-race antiface, and the non-central antiface is 17% more similar to its reference than the mixed-race antiface to the mixed-race average. We therefore multiplied the (already adjusted, see above) AEs, again, by 1.04 (same-race trajectory) and 1.17 (non-central trajectory).

Then, the effect of trajectory was re-examined in a two-way ANOVA with trajectory as repeated measures factor and face race as a between-participants factor. The ANOVA revealed a significant main effect of trajectory [$F(2,50) = 4.323$, $p = 0.019$, partial $\eta^2 = 0.147$], but no main effect of face race [$F(1,25) = 1.443$, $p = 0.241$, partial $\eta^2 = 0.055$] and no interaction [$F(2,50) = 0.557$, $p = 0.577$, partial $\eta^2 = 0.022$]. Pair-wise t-tests show that aftereffects remained larger for same than for mixed and non-central trajectories [*same vs. mixed:* $t(26) = 2.523$, $p = 0.018$; *same vs. non-central:* $t(26) = 2.919$, $p = 0.007$], whereas the aftereffects for mixed and non-central trajectories were not significantly different from each other [$t(26) = 0.110$, $p = 0.913$]. See Table 2 for means and SEs.

Table 2. Mean (SE) similarity of adapting **antifaces** to same-race and mixed-race **reference faces** (-50% vs. 0%)

	Reference / Trajectory		
Face Race	same-race	mixed-race	non-central
Asian	5.35 (0.28)	4.93 (0.24)	6.10 (0.20)
Caucasian	5.31 (0.29)	5.32 (0.22)	6.31 (0.30)
All	5.33 (0.20)	5.13 (0.17)	6.21 (0.17)

General Discussion and Outlook

The aim of this thesis was to examine several aspects of how faces are looked at, processed, perceived and represented in the brain. This aim was pursued through a series of studies that used different approaches and methods to answer the respective research questions. In the following, I will summarize the main findings of each of these studies and further discuss them in the context of current research. I will also outline potential ideas for future work or ongoing studies that are based on the results of this dissertation.

Symmetry in face representations

With respect to the first question on eye-movements in human face perception, an experiment was conducted where observers' viewing behavior was recorded while they compared pairs of face stimuli (**Chapter 1**, *pp 21-60*). Apart from a task-dependent pattern of visual exploration of the stimuli, the study revealed, unexpectedly, that faces were compared in a mostly asymmetrical way, that is, only the inner halves of both faces in a pair were thoroughly examined. This suggests that our participants considered the faces they were presented with as symmetrical, although we know from everyday experience with faces that this is not always the case. Unless somebody's face is extremely regular, we are sometimes surprised to see ourselves on a photograph, that is, from an inverse viewpoint than the face we see in the mirror every morning. The first interpretation of the results was thus that our participants kept their attention to the inner halves of the two faces in a pair simply for reasons of efficiency, or that they were concerned about the eye-tracking accuracy and therefore did not dare to move their head any more than necessary. However, as described in the discussion section of **Chapter 1** (*pp 52-60*), the results remained the same even when participants were allowed to take as much time as needed (in the pilot experiment) or were free to move their head a little (in a control study). One could of course also claim that the "inner features" scanning pattern just comes from a "center bias" often observed in perception experiments where images are presented on a computer screen (e.g., Buswell, 1935; Mannan, Ruddock, & Wooding, 1995, 1996, 1997; Bindemann, 2010). This bias does not depend on the location of the fixation marker preceding image onset (Bindemann, Scheepers, Ferguson, & Burton, in press; Tatler, 2007) and does not reflect a viewing preference for a straight-ahead position (Vitu, Kapoula, Lancelin, & Lavigne, 2004) but rather a systematic bias towards the center of a visual display. However, the central viewing tendency determines predominantly *initial* fixation positions, and scene content of high relevance does nevertheless influence saccade landing positions (Bindemann, 2010, Bindemann, in press). In our experiment, where the center of the screen coincided with the fixation cross prior to stimulus onset, such a center bias would therefore not explain why observers restrained from looking at the outer halves of the face stimuli even *after* the first or the first few fixations. Especially when task difficulty is high and observers are instructed to move their head and take their time to gather the information needed to answer as correctly as possible, we would expect them to consciously search for information that they might have ignored before due to such an automatic bias. The inner-half bias might therefore reveal something about the way information from faces is represented in the brain, rather than it being a technical side-effect of our study. Maybe ignoring the actual symmetry or asymmetry of a face is just a very efficient approach of the visual system to deal with most day-to-day face perception tasks: Although people are highly sensitive to individual differences in facial symmetry (e.g., Rhodes et al. 1998; Perrett et al., 1999; Rhodes, Chan, Zebrowitz et al., 2003; Simmons et al. 2004), this information seems to be used primarily for judgments of attractiveness and other aspects of mate quality, such as health (Rhodes et al. 2001b; Zebrowitz & Rhodes 2004). Since a higher or lower degree of symmetry does not, however, seem to change the perception of the sex or race of a face, or how masculine, sympathetic, happy or communicative we find someone, this information might be disregarded if not essential in a given situation. It could be interesting to test this assumption by using one set of face stimuli with varying degrees of symmetry in an experiment where observers are asked to judge characteristics that either are or are not related to aspects of mate quality. Such a design would allow to directly determine for which of these judgments observers take into account face symmetry and for which they disregard this information even if it is available.

A common criticism of our approach of inferring the information entering the visual system from observers' point of fixation on a stimulus is that it is in fact possible to retrieve a lot of information from just one glance at a face. A visual stimulus can be explored by directly fixating single parts of it one after the other, but there is of course also information from peripheral vision, outside the very center of gaze. In a recent study, Hsiao and Cottrell (2008) demonstrated that face recognition (i.e. deciding whether a face has been previously seen) could indeed be achieved by making only one or two fixations to a central point on a face stimulus. This finding suggests that a brief "holistic" overview without detailed local facial information can be sufficient for making familiarity decisions. On the other hand, as mentioned before, Henderson and colleagues found that restricting the eye gaze of observers to the center of a (real-size) face during a learning phase clearly impairs later recognition performance of that face (Henderson, Williams, & Falk, 2005), suggesting that the "holistic glance" does not always provide enough information for encoding identity. Besides, there are studies on eye movements on faces demonstrating that prior experience with faces affects fixation patterns in a face recognition task – which would not be useful if one central fixation was enough to recognize somebody. In a study from de Belle and colleagues, for example, the local features of familiar faces were processed more than those of unfamiliar faces (van Belle, Ramon, Levebre, & Rossion, 2010).

It could be interesting to test which kind of judgments can be made on the basis of just one brief central fixation at a face stimulus (e.g., familiar or not, which age, male or female, happy or sad, etc.) and how decisions might change with longer free-viewing presentation times. However, to come back to the study presented in **Chapter 1** and the criticism that the point of fixation on a face does not accurately reveal which information is processed: It is important to note, again, that our goal here was *not* to find out which information in faces is useful or necessary to solve a certain task. Rather, we were interested in the scanning strategies that observers (consciously or not) seem to consider relevant to solve that task, since these strategies, independent of their appropriateness and effectiveness give insight into the way faces are encoded and represented in the brain.

Sex differences in eye movements to faces

Another important finding from the study presented in **Chapter 1** is that eye-movements of male and female participants differed from each other - but only in those conditions where the sex of the face stimuli was a relevant feature to solve the task (see Results section, *pp 34-51*). Since both groups of participants were equally good at discriminating between the faces, despite different viewing behavior, these intriguing results suggest that there is something different about the way male and female observers encode faces, in this case faces of different sexes, in their brain. As they paid attention to different features of the face stimuli, they might encode faces relative to slightly different dimensions, or weight these dimensions differently, at least regarding the sex of a face. Whether there are differences between both sexes regarding information processing in general is an ongoing question in behavioral and brain research.

Men and women seem to differ not only in their physical attributes and reproductive function but also in many other characteristics, including certain personality traits (e.g., Alan Feingold, 1994; Costa, Terracciano & McCrae, 2001) and the way they solve intellectual problems. According to the literature, females consistently perform better on tasks involving language production and comprehension, and fine motor skills; males on the other hand show better performance on tasks where visuospatial operations and fluid reasoning are involved (Beatty, 1984; Halpern,1997; Levy & Heller, 1992). Females have also been found to perform better on episodic memory tasks including delayed recall and recognition than males, but males and females do not show differences in working, immediate and semantic memory tasks (Halpern, 2000; Herlitz, Nilsson, & Backman, 1997; Herlitz, Airaksinen, & Nordstrom, 1999). Since mnemonic capacity per se does not differ between males and females (Maccoby & Jacklin,

1974), the sex differences in memory performance have been proposed to rather reflect underlying differences in the strategies used to process information (McGivern et al., 1997; Meyers-Levy, 1989).

As to human face perception, females in general outperform men in recognizing faces (Lewin, Wolgers & Herlitz, 2001; Cellerino et al., 2004; Rehnmann & Herlitz, 2006), largely due to superiority in recognizing female faces (McKelvie, 1987; McKelvie, Standing, St. Jean, & Law, 1993; Lewin & Herlitz, 2002). Women and girls, according to previous research, exhibit an "own-sex bias", i.e., they perform at a higher level for female faces than for male faces (Cross, Cross, & Daly, 1971; Lewin & Herlitz, 2002; Sugisaki & Brown, 1916; Temple & Cornish, 1993; Wright & Sladden, 2003), while men and boys do not show a corresponding own-sex bias for male faces. In brief, these findings suggest that the perception of male and female faces differs for male and female observers, as did the eye-tracking results presented in the study in **Chapter 1**. Our finding that men and women attend to different aspects of the same visual stimuli could reflect pre-existing cognitive biases that either contribute to or result from sex differences in neural functioning. Differences in functional involvement of brain regions between men and women have been found throughout the brain, during, for example, the processing of language, memory, emotion, or vision (see e.g., Cahill, Uncapher, Kilpatrick, Alkire, Turner, 2004; Cahill, Gorski, Belcher, Huynh, 2004). While processing face stimuli, male subjects have been found to show an asymmetric functioning of visual cortex, while female observers showed a more bilateral functioning, using event-related potentials (Proverbio et al., 2006). A functional magnetic resonance imaging (fMRI) study on sex differences in brain responses in face-relevant areas demonstrated differential amygdala and anterior temporal cortex activation in men and women during exposure to male and female faces, with higher activation in men during exposure to female faces (Fischer et al., 2004). All these findings suggest that males and females differ in the cognitive strategies they use to process information in general, and especially facial information.

The results presented in **Chapter 1** offer an ideal starting point for further investigating the question of processing differences between male and female participants for human faces. As a follow-up study, we are currently testing whether the sex differences in eye movements we found remain present when observers judge other facial characteristics, apart from sex. Here, participants rate faces of different sexes (and races, to simultaneously explore the cross-race question, see below) regarding several characteristics that are consistently used in face perception research (for reasons of relevance), e.g., intelligence, attractiveness, or trustworthiness. An alternative future study would be to manipulate the task, instead of the characteristic under investigation. We found that male and female participants direct their attention to different facial features when dealing with the sex of faces, and the current eye tracking study will first show whether this is also the case for other characteristics. Then, introducing a face inversion task, for example, where the overall configuration of the stimulus is destroyed but all the visual information is still there, could show whether the different viewing patterns are due to the fact that one group of participants (i.e., males or females) relies less on featural facial information than the other, as suggested in studies in some other domains of visual perception (e.g., Guillem & Mograss, 2005).

Male and female faces in male and female brains

The perception of male and female faces was also explored in the study presented in **Chapter 2** (*pp 61-84*), where participants had to categorize and discriminate between faces of both sexes. As it seems, in spite of the social and biological relevance of the sex of a person, faces of different sexes are not perceived categorically. Consequently, not every face is classified as either male or female, but somebody's face can also be perceived as "ambiguous" concerning its sex. This seems surprising, considering that it is essential to know if we are talking to a man or a woman in a variety of social situations. Moreover, even today there are still specific gender roles in many societies we are supposed to more or less identify with. Most studies that have investigated categorical perception of human faces (regarding identity, sex, and race) have used stimuli that were morphed between different identities as well as between different sexes and races, and/or participants were exposed to the same identities over and over again (e.g., Levin &

Beale, 2000; Beale & Keil, 1994; Campanella, Hanoteau, Seron, et al., 2003; Campanella, Chrysochoos, & Bruyer, 2001). When familiarity was controlled, CP did not occur for the sex of the face stimuli (see Bülthoff & Newell, 2004, and **Chapter 2**). It might thus be possible that these categorical effects were rather due to familiarization with the faces in the course of the experiment, than to actual categorical perception of the race and sex of faces. As the perception of familiar face identities is indeed categorical (Beale & Keil, 1994) it would explain these findings, and also the lack of CP in the study presented in **Chapter 2**. Here, identity was held constant and familiarization was avoided, so that a categorical effect could have only arisen from the change in sex information alone – which it did not.

Another explanation could be, however, that the appearance of our stimuli, derived from 3D scans of real heads, is responsible for the lack of CP: They are rather natural-looking compared to 2D photographs, but are devoid of any external feature that could serve as a cue, like hair, glasses or make-up. It has been suggested, however, that unfamiliar faces are often processed in terms of these external cues (Bruce et al., 1999), and that when we get familiarized with a face (or a person), the focus moves more to inner facial features and these might then play a more important role in recognition (e.g., Ellis et al., 1979; O'Donnell & Bruce, 2001; Stacey, Walker, & Underwood, 2005). Maybe accurate sex judgments are also made on the basis of external rather than inner-face information, as long as a person is not familiar to us. Hair-style, make-up, or even cultural "accessories" like a head scarf would thus serve as the essential cue to sex. If we personally know somebody, on the other hand, we already have a clear representation of their sex. It would therefore make sense to not process this information independently of the identity of the person but to have a common representation of all relevant information concerning that person. Although it might seem rather obvious - and since I am not aware of any experimental study that has shown this - it could be worth testing the hypothesis that CP of sex would indeed arise for unfamiliar face continua when they contain "superficial" sex-specific features, like hair. A simple CP experiment could be done using the same kind of stimuli as in the study presented in **Chapter 2**, but including clearly sex-specific hair styles. Identity and familiarization should still be controlled for, and the hair shape that would be morphed from male to female would serve as the external cue that was missing in the experiment presented here.

Recent brain imaging data support the view that sex is not processed independently from face identities, by showing that neural responses specific for sex information only are surprisingly weak and distributed across the whole network usually involved in face processing (Kaul, Rees & Ishai, 2010). This suggests that a huge number of neurons that are involved in the perception of faces do process sex information, but it is not abstracted from individual faces and separately "stored". To further investigate this question of how sex and identity of faces are processed, we are using another powerful fMRI method that gives insight into the nature of information processing in the brain, that is, neuronal adaptation combined with an attention-controlling orthogonal task. Rotshtein and colleagues reported that by applying this method, they could distinguish between brain regions encoding the continuous physical manipulation of face stimuli morphed between two identities, and those encoding the perceived identity category change shown by psychophysical experiments (Rotshtein et al. 2005). Following the same approach, and using the psychophysical procedure that was used in the CP study in **Chapter 2**, we designed a study to determine which brain areas show sensitivity to physical or category changes in faces manipulated in their sex. Are these the same that have been shown for identity manipulations, as in the study by Rotshtein and colleagues? That would be more direct evidence of a direct linkage between the processing of sex and identity of faces (proposed e.g., by Rossion, 2002, Ganel & Goshen-Gottstein, 2002, Bülthoff & Newell, 2004) and the lack of an independent representation.

In addition, the brain imaging results of male and female observers will be compared to investigate whether the observed strategic differences of participants' eye movements from **Chapter 1** are reflected in activation patterns in the brain. Both men and women have to recognize and classify human faces constantly. Until now, there is no evidence that they differ in their capacity to extract sex-related facial information necessary for their social life. There were also no differences in performance in the eye-tracking study in **Chapter 1** or in the categorical perception of male and female faces in the study in **Chapter 2**. However, since I found differences in how male and female participants gather visual information in faces and what kind of information they consider diagnostic in my eye-tracking experiments, male and female observers might also represent the facial dimension "sex" differently in their brain. Eye movements could, for example, reflect a more holistic or more feature-based processing in male than female observers, and this processing differences could be visible in neuronal activation patterns. Rossion and his colleagues reported face-specific regions in the middle fusiform gyri showed more activation in the right hemisphere when matching whole faces than face parts, whereas this pattern of activity was reversed in the left homologous region (Rossion et al., 2001). Localizer scans using the same face stimuli but different tasks (one requiring holistic, the other feature-based processing), could provide reference activation patterns that can subsequently be compared to male and female brain activation from the main adaptation scan session to reveal potential differences in processing of facial information.

A male bias in face sex classification

The second intriguing finding from the study presented in **Chapter 2** is the consistent male bias that was observed when participants had to classify the face stimuli into male and female. When first presenting participants with nearly 200 randomly chosen faces from the database, this finding did not seem so surprising. Face stimuli devoid of hair and make-up do look less female than real faces and are even sometimes considered to be younger males rather than adult females. This is true for other databases with face stimuli as well (e.g., Davidenko, 2007). However, after manipulating and choosing stimuli for the main experiment on the basis of the initial ratings, with the specific goal to get rid of the bias by providing a symmetrical range from male to female, we did not expect the effect to still be present. Interestingly, the same male bias was reported in a study using point-light walkers to induce adaptation aftereffects (Troje et al., 2006): In this study, observers also reported seeing more male than female walkers in a set of stimuli before adaptation, and even though the authors then choose their stimuli to control for this bias and keep the "sex range" of the walkers symmetrical, the bias remained almost constant. This and our finding suggest that there is a perceptual or cognitive bias to answer "male" when one is not absolutely sure about a person's sex. In the case of human faces, this interesting phenomenon has been suggested to result from the fact that anatomically, there are no distinctly "female features" (e.g., Enlow, 1990) in faces in general. The answer "female" when classifying a face's sex would thus mean "there are no male traits".

In accordance with what I hypothesized above about the lack of CP for the sex of unfamiliar faces, the "no male traits" interpretation of the male bias suggests that external features, i.e., hairstyle, makeup, clothing, maybe even behavior, might be used as cues to a person's sex, more than just physical appearance of a person or of their face itself. Body shape can of course be a very salient cue, but even if it is not available, or not very informative due to ambiguous or disguising clothing, we still need to make a decision. Interestingly, all the then remaining external features are defined and shaped by culture and thus are not biologically "hardwired" and universal. While long hair is still a very salient cue that spontaneously biases one to classify a person as female in most Western societies, covered hair would be the quickest and most prominent female feature in many strictly Islamic countries. As to the whole body appearance, trousers used to be and still are in some societies a strictly mal-specific piece of cloth, which then makes a skirt or dress the "no male traits" signal. A rather speculative but nevertheless possible explanation for such a (culturally defined) bias could be that misclassifying a male person as female has generally proved to be potentially more dangerous than misclassifying a woman as a man in the history of humans.

Norms and face spaces

Finally, **Chapter 3** (*pp 85-114*) dealt with the question how the widely accepted multidimensional face space model might be organized, more specifically with respect to faces of different races. Here, Asian and Caucasian faces were used in a high-level adaptation paradigm, to reveal how these two face categories are represented in the brain. The results suggest that a face's identity is encoded relative to a race-specific norm face, rather than a general norm representing faces of all races.

These results seem very plausible as an explanation for the "other race effect" (ORE). The ORE phenomenon is based on the finding that faces are easily and quickly classified by ethnicity (e.g., Levin, 1996), but that we are not efficient at perceiving and encoding differences between people whose faces do not share the morphological features we are used to see in our everyday live (e.g., Bothwell et al. 1989, for a recent meta-analysis see Meissner & Brigham, 2001). It suggests that our face processing system has been tuned very finely through visual experience: The mechanisms developed to recognize individual faces of our own race cannot be generalized efficiently to faces of another race. This is not surprising if one considers that face expertise needs long experience and that adult face processing abilities are not fully developed until adolescence, as shown by a number of studies in face processing development (e.g., Schwarzer, 1997; Itier & Taylor, 2004). Valentine's framework explained the other-race phenomenon in terms of face space dimensions that are encoded during learning of the features relevant to recognize the population of faces we deal with regularly, and which end up as inappropriate for discriminating between faces from other races that are characterized by different "typical" features. Consequently, other-race faces, according to Valentine's framework, build a dense cluster distant from the center of the environmentally defined face space, which makes them more confusable with each other (which is what the ORE predicts) but on the other hand leads to very fast race classification, faster than judgments of other traits like age or sex (Montepare & Opeyo, 2002). The "cluster view" is in agreement with the finding presented in **Chapter 3**, that is, dissociable norms for faces of different races that are used to encode each individual face identity according to its own race category. Such norms, built up and also shifted and fine-tuned over time (as becomes evident in face adaptation experiments, e.g., Leopold et al., 2001; Webster, Kaping, Mizokami, et al., 2004), provide a very intuitive explanation for the other-race effect: The norm that is built-up over a lifetime of exposure to an infinite number of individuals is of course more fine-tuned, to discriminate between hundreds or even thousands of different people, than the norm that is calculated over a small number of faces, because one only knows very few people from unfamiliar races on a personal level. As said above, children's performance on face perception tests does not reach adult levels until adolescence, a result which, a priori, could be due to qualitative changes in face mechanisms with age. Research has established, however, that face-space coding is already norm-based in children younger than 8 years of age (Nishimura, et al., 2008; Jeffery et al., 2010). Since 8-year-olds do not yet behave like adults in judging identity though, based on the spatial relations of facial features or across changes in viewpoint (e.g., Mondloch, Le Grand, & Maurer, 2002, 2003), it is very well possible that their face space dimensions are not yet well defined, in accordance with the finding that 8-year-olds tend to rely on a few salient features when making perceptual judgments on faces (Schwarzer, 2000; Nishimura et al., 2009). Accordingly, fine-tuning of the own-race norm and its appropriate dimensions (but not of useful dimensions for other more unfamiliar races) might be the source of developmental improvement in face identification performance beyond preschool age – and also the reason for less good memory and discrimination abilities for other-race faces.

Faces of one's own race seem to be perceived more holistically than faces of unfamiliar races (Michel et al., 2006), and this finding has been suggested to be a result of different levels of expertise with these groups of faces and thus possibly, in the context of the face space framework, of less well defined dimensions for other-race faces. Also very interestingly, data from human and macaque eye-movements on human and macaque faces show that fixation patterns are critically affected by presenting either conspecific or nonconspecific faces (Dahl, Wallraven, Bülthoff, & Logothetis, 2009). When face stimuli were blurred or inverted (i.e., either featural or configural information present in the stimuli was manipulated) eye-movements of both species were modulated in a similar, systematic way. The authors attribute this finding also to an effect of perceptual expertise, given that macaques are experts for macaque faces and humans are experts for human faces, which might again be a sign of dimensions that are well defined for the known and less well defined for the unknown group of faces - in this case a rather unknown species. Coming back to the level of human race, the eye-movement study described above about differences in scanning patterns of male and female observers when making various judgments on faces was designed to simultaneously explore the question of scanning differences between observers of different races. To that end, the experiment is being run at the Max Planck Institute in Tübingen, Germany and simultaneously at the Department of Brain and Cognitive Engineering, Korea University, Seoul, Korea. Thus observers are male and female and of either Caucasian (German or Western European) or Asian (Korean) origin. Stimuli include faces of different sexes as well as races, so that all possible combinations of observers and faces can be compared to gain insight into differences in processing information from male or female faces of one's own versus an unfamiliar race. If the characteristics that are judged (i.e., intelligence, trustworthiness, etc.) represent dimensions relevant for encoding of and discriminating between faces, then according to observers' varying degree of expertise with Asian and Caucasian faces, face space dimensions for these characteristics might be defined differently. This could then, according to what has been shown for humans' and macaques' scanning patterns on face stimuli, become evident in distinct exploration patterns of both groups of face stimuli in Asian and Caucasian observers.

It has been shown that the averageness of a face corresponds to its perceived attractiveness (e.g., Valentine et al. 2004, Rhodes et al., 2000; Rhodes et al., 2003), that is, the closer a face is to the average of the face space, the more attractive it is. If Valentines assumption of an other-race cluster distant from the face space center was correct, and following the averageness-attractiveness relationship, then one may ask whether other-race faces (as they are distant from the average face) should not be perceived as less attractive than own-race faces (which are closer to the average). Western Europe and the US have of course a long history of ethnocentrism and "white supremacy", and it has been reported that racist attitudes in Caucasian subjects lead to rating black faces as less attractive than white faces (Fazio, Jackson, Dunton, & Williams, 1995). However, such a bias is not generally found nowadays, as for example Rhodes and colleagues (Rhodes et al., 2005) showed for attractiveness ratings of Asian faces. Here again, the results from the study presented in **Chapter 3** provide an easy explanation for such an inconsistency: Other-race faces are not encoded far away from the general norm of the space, but distant from the average of the familiar own-race face population, in relation to a mean of their own. As a consequence, we judge faces of unfamiliar races as more or less attractive depending on how close they are to their own norm, independently of the norm for faces of familiar races.

Given the results above, the question remains how norms to represent faces relative to different subcategories are built up over time. More specifically, the findings raise questions that relate to the (typical and impaired) development of the perception of faces of different races over time and from an early age onwards. It would be very interesting to find out how personal experience with different races leads to specifically shaped norms in each individual. A way to correlate personal experience with norm development over time could be to test children of different ages, from different ethnic backgrounds and children with developmental abnormalities affecting their face perception abilities, using high-level face adaptation paradigms as the one described here. This could also shed some light on the question how the process of norm development with exposure and age influences the robust and reliable perception of humans as belonging to different race groups. This phenomenon is in fact surprising, given that by the criteria biologists typically use to apply the concept of "subspecies" or "races", humans do not qualify (see for example Cosmides et al., 2003). Nevertheless, as mentioned above, other-race faces seem to be processed differently than own-race faces (Michel et al., 2006), and using a pattern-classification approach, Natu and colleagues (Natu, Raboy, & O'Toole, 2010) showed that Asian and Caucasian faces elicit dissociable neural responses in the brain. Given the social relevance of race classification, not least when it comes to stereotyping and prejudice, understanding the processes that shape our face-space and thus our perception of other people in general is a crucial issue, possibly also in view of education.

A norm-based face space framework is an elegant and economical way to represent faces. By coding not every single identity but rather how every face deviates from a norm that represents the central tendency of a distribution of faces, the visual system can see past the shared and highly redundant structure of faces and easily code those subtle variations that define individuals. Calculating the norm as an average over a distribution of faces also allows flexible updating in response to changes in the statistical distribution of dimension values in the population of faces encountered over time. This idea naturally raises the question how such a space could be implemented in the brain. A simple neural model has been proposed (Rhodes et al. 2005; Rhodes & Jeffery 2006; Robbins et al. 2007; Tsao & Freiwald 2006), similar to those used for norm-based coding of simpler properties like direction of motion and aspect ratio (Mather 1980; Regan & Hamstra 1982; Sutherland 1961; Suzuki 2005). This model assumes that each dimension of face-space is coded by a pair of neural populations, with one population coding above-average values (e.g., large eyes) and the other one coding below-average values (e.g., small eyes). There is little response to average inputs which are signaled by equal activation in both populations, so that neural responses are tuned to distinctive inputs (in accordance with human neuroimaging and monkey neurophysiology data showing low activation to an average face and increasing activation with distance from the average, Leopold et al., 2006; Loffler et al., 2005). The result is a very economical system that codes the norm only implicitly, in the tuning functions of the populations coding each dimension.

In this thesis, the perception and encoding of mainly two characteristics of human faces was investigated, i.e., sex and race. Hypothesizing about how the results of the studies presented here fit into current models of a face space representation in the brain leads us back to the question that was brought up in the introduction, i.e., what are actually the dimensions used to encode faces in such a space? Valentine's framework did not, as said before, specify the nature and number of dimensions necessary to represent faces most efficiently; Valentine proposed only that the dimensions represent cues to identity that are important for discriminating faces. A dimension might thus represent a very obvious, easily identifiable feature (e.g., eye color), or it might be a more complex combination of cues that cannot be verbalized easily. Few attempts have been made to number or label the dimensions of this space, and it is indeed not a trivial problem. For example, manipulating face stimuli to test whether observers are able to perceive and use the differences introduced between them in that way requires a priory definition of the facial characteristics to be tested; it does not reveal face dimensions that the experimenter has not thought of before. Moreover, when observers are asked to discriminate between faces differing in a certain characteristic, they might usually be able to do so – this does, however, not proof they use the available information also to encode faces in their mental face space representation. Measures of physical differences between groups of different face categories (e.g., male-female, attractive-ugly, etc.) can be derived from face databases, but again, these measures do not reveal the mental face space representation but the variation present in the stimulus set. One approach to describing the dimensions of face-space is to use multidimensional scaling (MDS), a statistical procedure that represents measurements of perceived similarity among pairs of objects as distances between points in a multidimensional space. MDS is often able to show regularities that can not be derived directly from raw similarity judgments (Borg & Groenen, 2005); however, it is sometimes difficult to interpret the space dimensions found in this way. Also, the variations provided within the stimulus set predefine the outcome. Shepherd, Ellis and Davies (1977) used MDS techniques to demonstrate that face shape, hair length, and perceived age were potentially significant dimensions for face space. An MDS applied to similarity ratings on male Caucasian faces with similar hair styles - thus in this case, hair length was excluded as a potential dimension - revealed that the most useful dimensions appeared to represent face width, perceived age, facial hair, and forehead size (Johnston, Milne, Williams, & Hosie, 1997). Busey (1998) found six interpretable dimensions when using MDS on bold male face stimuli of different races, and labeled them age, race, facial adiposity, facial hair, aspect ratio of head, and color of facial hair. Using a stimulus set including faces of different sexes would thus be very likely to reveal sex as one of the first dimensions, maybe at the expense of another dimension that would only become visible within each sex category, that is, at a finer scale of differences between faces. Nevertheless, when preselection of stimuli is well-controlled regarding dimensions of face space that are either obvious or have been demonstrated to be of relevance, MDS provides a very useful tool to quantify the psychological face space or potential "subspaces" of it. Testing observers of different races and sexes could give insight into different weighting of dimensions for face space representations in these groups.

Using, on the other hand, several stimuli sets each consisting of faces of only one single race, or one sex, or faces of the same age, would allow comparing directly across these subcategories of faces and thus reveal how the dimensions to encode them differ.

Afterthought

The research described in this thesis aims to contribute towards a better understanding of how we look at faces, how they are perceived and encoded in the brain. Different methods and approaches were employed, each of them offering specific possibilities and giving insights into various aspects of face perception and processing. In a sense, it might seem as if this work has raised more questions than it is able to answer. However, a doctoral thesis is not meant to simply be a realized and completed scientific project that expands our knowledge about the world. Rather, it could also be taken as a starting point for a variety of potential future avenues to follow, built on the many ideas that went into the work and the results and experiences drawn from it. Measured against this scale, I suppose that my dissertation has served its purpose.

References

Alley, T.R., & Schultheis, J.A. (2001). Is facial skin tone sufficient to produce a cross-racial identitfication effect? *Perceptual and Motor Skills* 92(3), 1191-1189.

Angeli, A., Davidoff, J., & Valentine, T. (2001). Distinctiveness induces categorical perception of unfamilar faces. *Perception*, 30(Suppl.), 58.

Anstis, S.M. (1974). Chart demonstrating variations in acuity with retinal position. *Vision Research* 14(7), 589-592.

Armann, R., Bülthoff, I. (2009). Gaze behavior in face comparison: The roles of sex, task, and symmetry. *Attention, Perception, & Psychophysics* 71(5), 1107-1126.

Barlow, H.B., & Hill, R.M. (1963). Evidence for a physiological explanation of waterfall phenomenon and figural aftereffects. *Nature* 200(491), 1345.

Barton, J.J.S., Radcliffe, N., Cherkasova, M.V., Edelman, J., & Intriligator, J.M. (2006). Information processing during face recognition: The effects of familiarity, inversion, and morphing on scanning fixations. *Perception*, 35, 1089-1105.

Baxter, J.C. (1970). Interpersonal spacing in natural settings. *Sociometry*, 33, 444-456.

Beale, J.M., & Keil, F.C. (1995). Categorical effects in the perception of faces. *Cognition*, 57(3), 217-239.

Beatty, W.W. (1984). Hormonal organization of sex-differences in play fighting and spatial-behavior. *Progress in Brain Research* 61, 315-330.

Benson, P.J., & Perrett, D.I. (1994). Visual processing of facial distinctiveness. *Perception* 23, 75-93.

Bestelmeyer, P.E.G., Jones, B.C., & DeBruine, L.M. (2008). Sex-contingent face aftereffects depend on perceptual category rather than structural coding. *Cognition* 107, 353-365.

Bindemann, M. (2010). *Scene* and *screen* bias early eye movements in scene viewing.

Vision Research 50, 2577-2587.

Blanz, V. (2000). Automatische Rekonstruktion der dreidimensionalen Form von Gesichtern aus einem Einzelbild. Dissertation zur Erlangung des Grades eines Doktors der Naturwissenschaften, Eberhard-Karls-Universitaet Tuebingen.

Blanz, V., & Vetter, Th. (1999). A Morphable Model For The Synthesis Of 3D Faces. In *Proceedings of SIGGRAPH 99*, 187–194.

Bornstein, M.H., & Korda, N.O. (1984) Discrimination and matching within and between hues measured by reaction times – some implications for categorical perception and levels of information-processing. *Psychological Research-Psychologische Forschung* 46(3), 207-222.

Bothwell, R.K., Brigham, J.C., & Malpass, R.S. (1989). Cross-racial identification. *Personality and Social Psychology Bulletin* 15(1), 19-25.

Boutet, I., Collin, C., & Faubert, J. (2003). Configural face encoding and spatial frequency information. *Perception & Psychophysics* 65(7), 1078-1093.

Brown, E., & Perret, D.I. (1993). What gives a face its gender? *Perception* 22(7), 829-840.

Bruce, V., & Young, A. (1986). Understanding face recognition. *The British Journal of Psychology*, 77(3), 305-327.

Bruce, V., & Young, A. (1998). In the Eye of the Beholder: The Science of Face Perception. New York: *Oxford University Press.*

Bruce, V., Burton, A.M., Hanna, E., Healey, P., Mason, O., Coombes, A., Fright, R., & Linney, A. (1993). Sex-discrimination – How do we tell the difference between male and female faces. *Perception* 22(2), 131-152.

Bruce, V., Doyle, T., Dench, N., & Burton, M. (1991). Remembering facial configurations. *Cognition* 38, 109–144.

Bruce, V., Langton, S., & Hill, H. (1999). Complexities of face perception and categorisation. *Behavioral and Brain Sciences* 22(3), 369.

Bruyer, R., Galvez, C., & Prairial, C. (1993). Effect of Disorientation on Visual Analysis, Familiarity Decision and Semantic Decision on Faces. *British Journal of Psychology*, 84, 433-441.

Bülthoff, I., & Newell, F. N. (2004). Categorical perception of sex occurs in familiar but not unfamiliar faces. *Visual Cognition*, 11(7), 823-855.

Burns, E.M., & Ward, W.D. (1978). Categorical perception – phenomenon or epiphenomenon – evidence from experiments in perception of melodic musical intervals. *Journal of the Acoustical Society of America* 63(2), 456-468.

Burton, A.M., & Vokey, J.R. (1998). The Face-Space Typicality Paradox: Understanding the Face-Space Metaphor. *The Quarterly Journal of Experimental Psychology* 51A(3), 475-483.

Buswell, G.T. (1935). How People Look At Pictures. Chicago, IL: *University of Chicago Press.*

Buswell, G.T. (1935). How people look at pictures: A study of the psychology of perception of art. *Chicago: University of Chicago Press.*

Butler, S., Gilchrist, I.D., Burt, D. M., Perrett, D. I., Jones, E., & Harvey, M. (2005). Are the perceptual biases found in chimeric face processing reflected in eye-movement patterns? *Neuropsychologia*, 43(1), 52-59.

Byatt, G., & Rhodes, G. (1998). Recognition of own-race and other-race caricatures: Implications for models of face recognition. *Vision Research* 38, 2455-2468.

Cahill, L., Gorski, L., Belcher, A., & Huynh, Q., (2004). The influence of sex versus sex-related traits on long-term memory for gist and detail from an emotional story. *Consciousness and Cognition* 13(2), 391-400.

Cahill, L., Uncapher, M., Kilpatrick, L., Alkire, M.T., & Turner, J. (2004). Sex-related hemispheric lateralization of amygdala function in emotionally influenced memory: An fMRI investigation. *Learning & Memory* 11(3), 261-266.

Calder, A.J., & Young, A.W. (2005). Understanding the recognition of facial identity and facial expression. *Nature Reviews Neuroscience* 6(8), 641-651.

Calder, A.J., Young, A.W., Benson, P.J., & Perrett, D. (1996). Self priming from distinctive and caricatured faces. *British Journal of Psychology* 87, 141-162.

Campanella, S., Chrysochoos, A., & Bruyer, R. (2001). Categorical perception of facial gender information: Behavioural evidence and the face-space metaphor. *Visual Cognition* 8(2), 237-262.

Campanella, S., Hanoteau, C., Seron, X., Joassin, F., & Bruyer, R. (2003). Categorical perception of unfamiliar facial identities, the face-space metaphor, and the morphing technique. *Visual Cognition* 10(2), 129-156.

Carbon, C.C., & Leder, H. (2006). Last but not least - The Mona Lisa effect: is “our” Lisa fame or fake?, *Perception* 35, 411-414.

Carbon, C.C., Strobach, T., Langton, S.R.H., Harsanyi, G., Leder, H., & Kovacs, G. (2007). Adaptation effects of highly familiar faces: Immediate and long lasting, *Memory & Cognition* 35, 1966-1976.

Cellerino, A., Borghetti, D., Sartucci, C. (2004). Sex differences in face gender recognition in humans. *Brain Research Bulletin* 63(6), 443-449.

Chen, J., Yang, H., Wang, A., & Fang, F. (2010). Perceptual consequences of face viewpoint adaptation: Face viewpoint aftereffect, changes of differential sensitivity to face view, and their relationship. *Journal of Vision, 10*(3):12, 1-11.

Christman, S. D., & Hackworth, M. D. (1993). Equivalent Perceptual Asymmetries for Free Viewing of Positive and Negative Emotional Expressions in Chimeric Faces. *Neuropsychologia*, 31(6), 621-624.

Chronicle, E. P., Chan, M. Y., & Hawkings, C. (1995). You can tell by the nose – Judging sex from an isolated facial feature. *Perception*, 24(8), 969-973.

Clifford, C.W.G. (2002). Adaptation-induced plasticity in perception: Motion parallels orientation. *Trends in Cognitive Sciences*, 6, 136-143.

Clifford, C.W.G., & Rhodes, G., Editors. (2005). Fitting the mind to the world: Adaptation and after-effects in high-level vision, Oxford University Press, Oxford.

Cosmides, L., Tooby, J., & Kurzban, R. (2003). Perceptions of race. *TRENDS in Cognitove Sciences* 7(4), 173-179.

Costa, P.T., Terracciano, A., & McCrae, R.R. (2001). Gender differences in personality traits across cultures: Robust and surprising findings. *2nd Annual Meeting of the Society for Personality and Social Psychology* 81(2), 322-331.

Cross, J.F., Cross, J., & Daly, J. (1971). Sex, Race and Beauty as Factors in Recognition of Faces. *Perception & Psychophysics* 10(6), 393.

Dahl, C.D., Wallraven, C., Bülthoff, H.H., & Logothetis, N. (2009). Humans and Macaques Employ similar Face-Processing Strategies. *Current Biology* 19, 509-513.

Davidenko, N. (2007). Silhouetted face profiles: A new methodology for face perception research. *Journal of Vision* 7(4)

Deffenbacher, K. A., Hendrickson, C., O'Toole, A. J., Huff, D. P., & Abdi, H. (1998). Manipulating face gender: Effects on categorization and recognition judgements. *Journal of Biological Systems*, 6(3), 219-239.

Ellis, A.W., Burton, A.M., Young, A., & Flude, B.M. (1997). Repetition priming between parts and wholes: Tests of a computational model of familiar face recognition. *British Journal of Psychology* 88, 579-608.

Ellis, H.D., Shepherd, J.W., & Davies, G.M. (1979). Identification of familiar and unfamiliar faces from internal and external features – some implications for theories of face recognition. *Perception* 8(4), 431-439.

Enlow, D.H. (1990). The plan of the human face. In: Dyson J. (ed.) Facial Growth, 3rd edn. W.B. Saunders Co., Philadelphia, 164-192.

Etcoff, N.L., & Magee, J.J. (1992). Categorical perception of facial expressions. *Cognition* 44(3), 227-240.

Falk, R.J., Hollingworth, A., Henderson, J.M., Mahadevan, S., & Dyer, F.C. (2000). Eye movements in human face learning and recognition. *Proceedings of the twenty-second annual conference of the Cognitive Science Society*, 1026

Farah, M. J., Wilson, K. D., Drain, M., & Tanaka, J. N. (1998). What is "special" about face perception? *Psychological Review*, 105(3), 482-498.

Feingold, A. (1994). Gender Differences in Personality – a Metaanalysis. *Psychological Bulletin* 116(3), 429-456.

Fischer, H., Sandblom, J., Herlitz, A., Fransson, P., Wright, C.I., & Bäckman, L. (2004). Sex-differential brain activation during exposure to female and male faces. *Neuroreport* 15(2), 235-238.

Frisby, J. P. (1980). Seeing: Illusion, mind and brain. Oxford: OUP.Hurlburt 2000;

Galpin A. J., & Underwood, G. (2005). Eye movements during search and detection in comparative visual search. *Perception & Psychophysics*, 67 (8), 1313-1331.

Ganel, T., & Goshen-Gottstein, Y. (2002). Perceptual integrality of sex and identity of faces: further evidence for the single-route hypothesis. *Journal of Experimental Psychology: Human Perception & Performance,* 28, 854–86.

Gauthier, I., Behrmann, M., & Tarr, M.J. (1999). Can face recongition really be dissociated from object recognition? *Journal of Cognitive Neuroscience* 11(4), 349-370.

Gilbert, C., & Bakan, P. (1973). Visual Asymmetry in Perception of Faces. *Neuropsychologia*, 11(3), 355-362.

Gitelman, D. R. (2002). ILAB: A program for post-experimental eye movement analysis. *Behavior Research Methods, Instruments, & Computers*, 34(4), 605-612.

Goffaux, V., Hault, B., Michel, C., Vuong, Q. C., Rossion, B. (2005).The respective role of low and high spatial frequencies in supporting configural and featural processing of faces. *Perception*, 34, 77- 86.

Goldberg, J. H., & Kotval, X. P. (1999). Computer interface evaluation using eye movements: methods and constructs. *International Journal of Industrial Ergonomics*, 24, 631-645.

Gosselin, F., & Schyns, P. G. (2001). Bubbles: a technique to reveal the use of information in recognition tasks. *Vision Research*, 41(17), 2261-2271.

Grill-Spector, K., Knouf, N., & Kanwisher, N. (2004). The fusiform face area subserves face perception, not generic within-category identitfication. *Nature Neuroscience* 7(5), 555-562.

Guillem, F., & Mograss, M. (2005). Gender differences in memory processing: Evidence from event-related potentials to faces. *Brain and Cognition,* 57(1), 84-92.

Guntekin, B., & Basar, E. (2007). Gender differences influence brain's beta oscillatory responses in recognition of facial expressions. *Neuroscience Letters,* 424, 94-99.

Halpern, D.F. (1997). Sex differences in intelligence - Implications for education. *American Psychologist* 52(10), 1091-1102.

Halpern, D.F., & LaMay, M.L. (2000). The smarter sex: A critical review of sex differences in intelligence. *Educational Psychology Review* 12(2), 229-246.

Harnad, S. (1987) Psychophysical and cognitive aspects of categorical perception: A critical overview. Chapter 1 of: Harnad, S. (ed.) (1987) Categorical Perception: The Groundwork of Cognition. New York: Cambridge University Press.

Havard C., & Burton A. M. (2006) The eye movement strategies performed during a face matching task. *Perception* 35, *210.*

Henderson, J. M., Williams, C. C., & Falk, R. J. (2005). Eye movements are functional during face learning. *Memory & Cognition,* 33(1), 98-106.

Henderson, J. M., Williams, C. C., Castelhano, M. S., & Falk, R. J. (2003). Eye movements and picture processing during recognition. *Perception & Psychophysics,* 65, 725-734.

Herlitz, A., Airaksinen, E., & Nordstrom, E. (1999). Sex differences in episodic memory: The impact of verbal and visuospatial ability. *Annual Meeting of the International Neuropsychology Society* 13(4), 590-597.

Herlitz, A., Nilsson, L.G., & Backman, L. (1997). Gender differences in episodic memory. *Memory & Cognition* 25(6), 801-811.

Hill, H., Bruce, V., & Akamatsu, S. (1995). Perceiving the sex and race of faces – the role of shape and color. *Proceedings of the Royal Society of London Series B – Biological Sciences* 261(1362), 367-373.

Hole, G.J. (1994). Configurational factors in the perception of unfamiliar faces. *Perception* 23(1), 65-74.

Hsiao, J.H.W., & Cottrell, G. (2008). Two fixations suffice in Face Recognition. *Psychological Science* 19(10), 998-1006.

Hsiao, J.H.W., Cottrell, G. (2008). Two Fixations Suffice in Face Recognition. *Psychological Science* 19(10), 998-1066.

Inn, D., Walden, K.J., & Solso, R.L. (1993). Facial prototype formation in children, *Bulletin of the Psychonomic Society* 31, 197–200.

Itier, R.J., Taylor, M.J. (2004). Effects of repetition and configural changes on the development of face recognition processes. *Developmental Science* 7(4), 469-487.

Jaquet, E., & Rhodes, G. (2008). Face aftereffects indicate dissociable, but not distinct, coding of male and female faces, *Journal of Experimental Psychology – Human Perception and Performance* 34, 101-112.

Jaquet, E., Rhodes, G., & Clifford, C.W.G. (2005) [The face distortion aftereffect and discrimination]. Unpublished raw data.

Jaquet, E., Rhodes, G., & Hayward, W.G. (2007). Opposite aftereffects for Chinese and Caucasian faces are selective for social category information and not just physical face differences, *Quarterly Journal of Experimental Psychology* 60, 1457-1467.

Jaquet, E., Rhodes, G., & Hayward, W.G. (2008). Race-contingent aftereffects suggest distinct perceptual norms for different race faces, *Visual Cognition* 16, 734-753.

Jeffery, L., McKone, E., Haynes, R., Firth, E., Pellicano, E., & Rhodes, G. (2010). Four-to six-year-old children use norm-based coding in face-space. *Vision Research* 50(10), 963-968.

Jenkins, R., Beaver, J. D., & Calder, A. J. (2006). I thought you were looking at me: Direction specific aftereffects in gaze perception. *Psychological Science*, 17, 506-513.

Kanwisher, N., McDermott, J., & Chun, M. M. (1997). The fusiform face area: A module in human extrastriate cortex specialized for face perception. *Journal of Neuroscience*, 17(11), 4302-4311.

Kaul, C., Rees, G., & Ishai, A. (2010). Perception of gender is a distributed attribute in the human face processing network. *Journal of Vision* 10(7), 707.

Kohn, A. (2007). Visual adaptation: Physiology, mechanisms, and functional benefits, *Journal of Neurophysiology* 97, 3155-3164.

Kranz, F., & Ishai, A. (2006). Face perception is modulated by sexual preference. *Current Biology*, 16(1), 63-68.

Lee, K., Byatt, G., & Rhodes, G. (2000). Caricature effects, distinctiveness and identification: testing the face-space framework, *Psychological Science* 11, 379–385.

Leopold, D.A., Bondar, I., & Giese, M.A. (2006). Norm-based face encoding by single neurons in the monkey inferotemporal cortex, *Nature* 442, 572–575.

Leopold, D.A., O'Toole, A.J., Vetter, T., & Blanz, V. (2001). Prototype-referenced shape encoding revealed by high-level aftereffects, *Nature Neuroscience* 4, 89–94.

Leopold, D.A., Rhodes, G. , Müller, K.-M., & Jeffery, L. (2005). The dynamics of visual adaptation to faces, *Proceedings of the Royal Society of London*, Series B 272, 897–904.

Levin, D. T. (1996). Classifying faces by race: The structure of face categories. *Journal of Experimental Psychology – Learning Memory and Cognition* 22(6), 1364-1382.

Levin, D.T., Beale, J.M. (2000). Categorical perception occurs in newly learned faces, other-race faces, and inverted faces. *Perception & Psychophysics* 62(2), 386-401.

Levy, J., & Heller, W. (1992). Gender differences in human neuropsychological function. *Handbook of Behavioral Neurobiology; Sexual differentiation,* 245-274.

Lewin, C., & Herlitz, A. (2002). Sex differences in face recognition – Women's faces make the difference. *Brain and Cognition*, 50(1), 121-128.

Lewin, C., Wolgers, G., & Herlitz, A. (2001). Sex differences favoring women in verbal but not in visuospatial episodic memory. *Neuropsychology* 15(2), 165-173.

Liberman A.M., Cooper, F.S., Shankwei, D.P., & Studdert, M. (1967). Perception of speech code. *Psychological Review* 74(6), 431.

Little, A.C., DeBruine, L.M., & Jones, B.C. (2005). Sex-contingent face aftereffects suggest distinct neural populations code male and female faces, *Proceedings of the Royal Society of London* Series B 272, 2283–2287.

Loffler, G., Yourganov, G., Wilkinson, F., & Wilson, H.R. (2005). fMRI evidence for the neural representation of faces, *Nature Neuroscience* 8, 1386–1390.

Loftus, G. R., & Mackworth, N. H. (1978). Cognitive determinants of fixation location during picture viewing. *Journal of Experimental Psychology: Human Perception and Performance*, 4(4), 565-572.

Maccoby, E.E., & Jacklin, C.N. (1974). Myth, reality, and shades of gray – what we know and don't know about sex differences. *Psychology Today* 8(7), 109-112.

MacLin, O.H. & Webster, M.A. (2001). Influence of adaptation on the perception of distortions in natural images. *Journal of Electronic Imaging,* 10, 100-109.

Macmillan, N.A., & Creelman, C.D. (1991). Detection theory: A user's guide. New York: *Cambridge University Press.*

Maddess, T., McCourt, M. E., Blakeslee, B., & Cunningham, R.B. (1988). Factors governing the adaptation of cells in Area-17 of the cat visual cortex. *Biological Cybernetics*, 59, 229-236.

Mannan, S., Ruddock, K.H., & Wooding, D.S. (1995). Automatic control f saccadic eye movements made in visual inspection of briefly presented 2-D images. *Spatial Vision* 9, 363-386.

Mannan, S., Ruddock, K.H., & Wooding, D.S. (1996). The relationship between the locations of spatial features and those of fixations made during visual examination of briefly presented images. *Spatial Vision* 10, 165-188.

Mannan, S., Ruddock, K.H., & Wooding, D.S. (1997). Fixation sequences made during visual examination of briefly presented images. ` 11, 157-178.

Mather, G. (1980). The movement aftereffect and a distribution-shift model for coding the direction of visual movement. *Perception* 9(4), 379-392.

Mather, G., Verstraten, F., & Anstis, S. (Eds.). The Motion Aftereffect: A Modern Perspective. Cambridge, Massachusetts: MIT Press (1998).

McGivern, R.F., Huston, J.P., Byrd, D. (1997). Sex differences in visual recognition memory: Support for a sex-related difference in attention in adults and children. *Brain and Cognition* 34(3), 323-336.

McKelvie, S.J. (1987). Sex-differences, lateral reversal, and pose as factors in recognition memory for photographs of faces. *Journal of General Psychology* 114(1), 13-37.

McKelvie, S.J., Standing, L., St.Jean, D., & Law, J. (1993). Gender differences in recognition memory for faces and cars – evidence for the interest hypothesis. *Bulletin of the Psychonomic Society* 31(5), 447-448.

McKone, E., Edwards, M., Robbins, R., & Anderson, R. (2005). The stickiness of face adaptation aftereffects, *Journal of Vision* 5, 822a.

Meissner, C.A., & Brigham, J.C. (2001). Thirty years of investigating the own-race bias in memory for faces - A meta-analytic review, *Psychology Public Policy and Law* 7, 3-35.

Mertens, I., Siegmund, H., & Grusser, O. J. (1993). Gaze Motor Asymmetries in the Perception of Faces during a Memory Task. *Neuropsychologia*, 31(9), 989-998.

Meyers--Levy, J. Gender differences in information processing: A selectivity interpretation. P. Cafferata & A. Tybout, (Eds) Cognitive and Affective Responses to Advertising, Lexington Books (1989).

Michel, C., Caldara, R., & Rossion, B. (2006). Same-race faces are perceived more holistically than other-race faces, *Visual Cognition* 14, 55-73.

Michel, C., Corneille, O., & Rossion, B. (2007). Race categorization modulates holistic face encoding, *Cognitive Science* 31, 911-924.

Michel, C., Rossion, B., Han, J., Chung, C. S., & Caldara, R. (2006). Holistic processing is finely tuned for faces of one' own race. *Psychological Science*, 17(7),

608-615.

Mondloch, C.J., Dobson, K.S., Parson, J. & Maurer, D. (2004). Why 8-year-olds can't tell the difference between Steve Martin and Paul Newman: Factors contributing to the slow development to the spacing of facial features. *Journal of Experimental Child Psychology* 89, 159-181.

Mondloch, C.J., Gerdart, S., Maurer, D., & Le Grand, R. (2003). Developmental changes in face processing skills. *Journal of Experimental Child Psychology* 86, 67-84.

Montepare, JM; Opeyo, A, 2002 The relative salience of physiognomic cues in differentiating faces: A methodological tool. *69th Annual Meeting of the Eastern-Psychological-Association* 26(1), 43-59.

Movshon, J.A., & Lennie, P. (1979). Pattern-selective adaptation in visual cortical neurons. *Nature*, 278, 573-599.

Natu, V., Raboy, D., & O'Toole, A. (2010). Differential spatial and temporal neural response patterns for own- and other-race faces. *Journal of Vision* 10(7), 692.

Ng, M., Boynton, G.M., & Fine, I. (2008). Face adaptation does not improve performance on search or discrimination tasks. *Journal of Vision* 8, 1-20.

Nishimura, M., Maurer, D., & Gao, X. (2009). Exploring children's face-space: A multidimensional scaling analysis of the mental representation of facial identity. *Journal of Experimental Child Psychology* 103, 355-375.

Nishimura, M., Maurer, D., Jeffery, L., Pellicano, E., & Rhodes, G. (2008). Fitting the child's mind to the world: Adaptive norm-based coding of facial identity in 8-year-olds. *Developmental Science* 11, 620-627.

O'Leary, A., & McMahon, M. (1991). Adaptation to form distortion of a familiar shape. *Perception & Psychophysics*, 49, 328-332.

O'Toole, A. J., Deffenbacher, K. A., Valentin, D., McKee, K., Huff, D., & Abdi, H. (1998). The perception of face gender: The role of stimulus structure in recognition and classification. *Memory & Cognition*, 26(1), 146-160.

O'Toole, A. J., Peterson, J., & Deffenbacher, K. A. (1996). An ‚other-race effect' for categorizing faces by sex. *Perception,* 25(6), 669-676.

O'Donnell, C., & Bruce, V. (2001). Familiarisation with faces selectively enhances sensitivity to changes made to the eyes. *Perception* 30(6), 755-764.

O'Toole, AJ; Vetter, T; Blanz, V. (1999). Three-dimensional shape and two-dimensional surface reflectance contributions to face recognition: an application of three-dimensional morphing. *Vision Research* 39(18), 3145-3155.

Pearson, A. M., Henderson, J. M., Schyns, P. G., & Gosselin, F. (2003). Task-Dependent Eye Movements During Face Perception. *Abstracts of the Psychonomic Society, 8*, 84.

Pellicano, E., Jeffery, L. , Burr, D., & Rhodes, G. (2007). Abnormal adaptive face-coding mechanisms in children with autism spectrum disorder, *Current Biology* 17, 1508-1512.

Perrett, D.I., Burt, D.M., Penton-Voak, I.S., Lee, K.J., Rowland, D.A., & Edwards, R. (1999). Symmetry and human facial attractiveness. *Evolution and Human Behavior* 20(5), 295-307.

Posner, M.I., & Keele, S.W. (1968). On genesis of abstract ideas. *Journal of Experimental Psychology* 77(3P1), 353.

Proverbio, A.M., Brignone, V., Matarazzo, S., Del Zotto, M., & Zani, A. (2006). Gender and parental status affect the visual cortical response to infant facial expression. *Neuropsychologia* 44(14), 2987-2999.

Regan, D., & Hamstra, S.J. (1982). Shape-discrimination and the judgement of perfect symmetry – dissociation of shape from size. *Vision research* 32(10), 1845-1864.

Rehnman, J., & Herlitz, A. (2006). Higher face recognition abilities in girls – Magnified by own-sex and own-ethnicity bias. *Memory*, 14, 289-296.

Rehnman, J., & Herlitz, A. (2007). Women remember more faces than men do. *Acta Psychologica*, 124(3), 344-355.

Rhodes G., & Jeffery, L. (2006). Adaptive norm-based coding of facial identity,

Vision Research 46, 2977–2987.

Rhodes, G. & Leopold, D.A. (2010). Adaptive norm-based coding of face identity. In A.J. Calder, G. Rhodes, M.H. Johnson, & J.V. Haxby (Eds), Handbook of Face Perception, Oxford: Oxford University Press (forthcoming).

Rhodes, G. (1996). Superportraits: Caricatures and recognition, *The Psychology Press, Hove.*

Rhodes, G., & Jaquet, E. (2010). Aftereffects reveal that adaptive face-coding mechanisms are selective for race and sex. In R.A. Adams Jr, N. Ambady, K. Nakayama & S. Shimojo (Eds). Social Vision. Oxford University Press: New York, in press.

Rhodes, G., & Jeffery, L. (2005). Norm-based coding of face identity, *Perception* 34, 319-340.

Rhodes, G., Brennan, S., & Carey, S. (1987). Identification and ratings of caricatures: implications for mental representations of faces, *Cognitive Psychology* 19, 473–497.

Rhodes, G., Carey, S., Byatt, G., & Proffitt, F. (1998). Coding spatial variations in faces and simple shapes: A test of two models. *Vision Research*, 38, 2307-2321.

Rhodes, G., Hickford, C., & Jeffery, L. (2000). Sex-typicality and attractiveness: Are supermale and superfemale faces super-attractive? *British Journal of Psychology* 91, 125-140.

Rhodes, G., Jeffery, L., Clifford, C.W.G., & Leopold, D.A. (2007). The timecourse of higher-level face aftereffects, *Vision Research* 47, 2291-2296.

Rhodes, G., Jeffery, L., Watson, T., Jaquet, E., Winkler, C., & Clifford, C.W.G. (2004). Orientation-contingent face aftereffects and implications for face coding mechanisms. *Current Biology* 14, 2119-2123.

Rhodes, G., Jeffery, L., Watson, T.L., Clifford, C.W.G., & Nakayama, K. (2003). Fitting the mind to the world: Face adaptation and attractiveness aftereffects, *Psychological Science* 14, 558–566.

Rhodes, G., Lee K., Palermo, R., Weiss, M., Yoshikawa, S., Clissa, P., Williams, T.,

Peters, M., Winkler, C., & Jeffery, L. (2005). Attractiveness of own-race, other-race, and mixed-race faces. *Perception* 34(3), 319-340.

Rhodes, G., Maloney, L.T., Turner, J., & Ewing, L. (2007). Adaptive face coding and discrimination around the average face, *Vision Research* 47 974-989.

Rhodes, G., Watson, T.L., Jeffery, L., & Clifford, C.W.G. (2010). Perceptual adaptation helps us identify faces, *Vision Research* 50, 963-968.

Rhodes, G., Zebrowitz, L.A., Clark, A., Kalick, S.M., Hightower, A., & McKay, R. (2001b). Do facial averageness and symmetry signal health? *Evolution and Human Behavior* 22(1), 31-46.

Robbins, R., McKone, E., & Edwards, M. (2007). Aftereffects for face attributes with different natural variability: Adaptor position effects and neural models, *Journal of Experimental Psychology: Human Perception and Performance* 33, 570-592.

Roberts, T., Bruce, V. (1988). Feature saliency in judging the sex and familiarity of faces. *Perception*, 17(4), 475-481.

Rosch, E., Mervis, C.B., Gray, W.D., Johnson, D.M., & Boyes-Braem, P. (1976). Basic objects in natural categories. *Cognitive Psychology* 8(3), 382-439.

Rossion, B. (2002). Is sex categorization from faces really parallel to face recognition? *Visual Cognition*, 9(8), 1003-1020.

Rossion, B., Schiltz, C., Robaye, L., Pirenne, D., & Crommelinck, M. (2001). How does the brain discriminate familiar and unfamiliar faces?: A PET study of face categorical perception. *Journal of Cognitive Neuroscience* 13(7), 1019-1034.

Rotshtein, P., Henson, R.N.A., Treves, A., Driver, J., & Dolan, R.J. (2005). Morphing Marilyn into Maggie dissociates physical and identity face representations in the brain. *Nature Neuroscience* 8(1), 107-113.

Rutherford, M.D., Chattha, H.M., & Krysko, K.M. (2008) The use of aftereffects in the study of relationships among emotion categories. *Journal of Experimental Psychology-Human Perception and Performance* 34(1), 27-40.

Sangrigoli, S., & de Schonen, S. (2004). Recognition of own-race and other-race faces by three-month-old infants, *Journal of Child Psychology and Psychiatry* 45,

1219-1227.

Schwaninger, A., Lobmaier, J.S., Wallraven, C., & Collishaw, S. (2009). Two Routes to face Peception: Evidence From Psychophysics and Computational Modeling. *Cognitive Science* 33, 1413-1440.

Schwarzer, G. (1997). Development of face categorization: The role of conceptual knowledge. *Sprache & Kognition* 16(1), 14-30.

Schwarzer, G. (2000). Development of face processing: The effect of face inversion. *Child development* 71, 391-401.

Schwarzer, G., Huber, S., & Dümmler, T. (2005). Gaze behavior in analytical and holistic face processing. *Memory & Cognition*, 33, 344-354.

Schyns, P. G., & Oliva, A. (1999). Dr. Angry and Mr. Smile: when categorization flexibly modifies the perception of faces in rapid visual presentations. *Cognition* 69, 243–265

Schyns, P. G., Bonnar, L., & Gosselin, F. (2002). SHOW ME THE FEATURES! Understanding Recognition From the Use of Visual Information. *Psychological Science*, 13(5), 402-409.

Shepherd, J., Davies, G., & Ellis, H. (1979). Relative Effectiveness of ratings and verbal descriptions in the recognition of faces. *Bulletin of the British Psychological Society* 32, 110

Sigala, R., Logothetis, N.K., & Rainer, G. Own-species bias in the representations of monkey and human face categories in the primate temporal lobe. *(submitted).*

Simmons, L.W., Rhodes, G., Peters, M., Koehler, N. (2004). Are human preferences for facial symmetry focused on signals of developmental instability? *Behavioral Ecology* 15(5) 864-871.

Sinha, P., Balas, B., Ostrovsky, Y., & Russel, R. (2006). Face recognition by humans : Nineteen results all computer vision researchers should know about. *Proceedings of the IEEE* 94(11), 1948-1962.

Stacey, P. C., Walker, S., & Underwood, J. D. M. (2005). Face processing and familiarity: Evidence from eye-movement data. *British Journal of Psychology*, 96,

1-17.

Sugisaki, Y., & Brown, W. (1916). The correlation between the sex of observers and the sex of pictures recognized. *Journal of Experimental Psychology* 1, 351-354.

Sutherland, N. S. (1961). Figural aftereffects and apparent size. *Quarterly Journal of Experimental Psychology,* 13, 222-228.

Suzuki, S. (2005). High-level pattern coding revealed by brief shape aftereffects. In C. W. G. Clifford & G. Rhodes (Eds.)., Fitting the mind to the world: Adaptation and aftereffects in high-level vision. *Oxford: Oxford University Press,* 135-172.

Tanaka, J. W., & Sengco, J. A. (1997). Features and their configuration in face recognition. *Memory & Cognition,* 25(5), 583-592.

Tanaka, J.W., & Farah, M.J. (1993). Parts and wholes in face recognition. *Quarterly Journal of Experimental Psychology Section A – Human Experimental Psychology* 46(2), 225-245.

Tatler, B.W. (2007). The central fixation bias in scene viewing: Selecting an optimal viewing position independently of motor biases and image feature distributions. *Journal of Vision* 7(14), 1-17.

Temple, C.M., & Cornish, K.M. (1993). Recognition memory for words and faces in schoolchildren – a female advantage for words. *British Journal of Developmental Psychology* 11, 421-426.

Troje, N.F., Sadr, J., & Geyer, H. (2006). Adaptation aftereffects in the perception of gender from biological motion. *Journal of Vision* 6(8), 850-857.

Tsao, D., Moeller, S., & Freiwald, W.A. (2008). Comparing face patch systems in macaques and humans. *Proceedings of the National Academy of Sciences of the United States of America* 105(49), 19514-19519.

Tsao, D., Sasaki, Y., Mandeville, J., Leite, F., Pasztor, E., Wald, L., Dale, A., Orban, G., Vanduffel, W., & Tootel, R. (2001). fMRI reveals face-selective activity in awake behaving macaque. *Society for Neuroscience Abstracts* 27(1), 320

Tsao, D.Y., & Freiwald, W.A. (2006). What's so special about the average face?

Trends in Cognitive Sciences, 10, 391-393.

Valentine, T. (1991). A unified account of the effects of distinctiveness, inversion, and race in face recognition, *Quarterly Journal of Experimental Psychology.*

Valentine, T. (1991). A Unified Account of the Effects of Distinctiveness, Inversion, and Race in Face Recognition. *The Quarterly Journal of Experimental Psychology* A, 43(2), 161-204.

Valentine, T. (2001). Face-space models of face recognition, In M.J. Wenger & J.T. Townsend (Eds). *Computational, geometric, and process perspectives on facial cognition: Contexts and challenges, Hillsdale, NJ: Erlbaum.*

Valentine, T., & Endo, M. (1992). Towards an exemplar model of face processing: the effects of race and distinctiveness. *The Quarterly Journal of Experimental Psychology Section A: Human Experimental Psychology*, 44(4), 671-703.

van Belle, G., Ramon, M., Levebre, P., & Rossion, B. (2010). Fixation patterns during recognition of personally familiar and unfamiliar faces. *Frontiers in Psychology* 1, 1-8.

Vetter, T; Poggio, T (1997) Linear object classes and image synthesis from a single example image. *IEEE Transactions on Pattern Analysis and Machine Intelligence* 19(7), 733-742.

Vitu, F., Kapoula, Z., Lancelin, D., & Lavigne, F. (2004). Eye movements in reading isolated words : Evidence for a strong bias towards the center of the screen. *Vision Research* 44, 321-338.

Viviani, P., Binda, P., & Borsato, T. (2007). Categorical perception of newly learned faces. *Visual Cognition* 15(4), 420-467.

Walton, G.E., & Bower, T.G.R. (1993). Newborns form "prototypes" in less than 1 minute, *Psychological Science* 4, 203–205.

Watson, T.L., & Clifford, C.W.G. (2003). Pulling faces: An investigation of the face-distortion aftereffect. *Perception*, 32, 1109-1116.

Webster, M.A., & MacLin, O.H. (1999). Figural aftereffects in the perception of faces. *Psychonomic Bulletin & Review*, 6, 647-653.

Webster, M.A., Kaping, D., Mizokami, Y. Duhamel, P. (2004). Adaptation to natural face categories, *Nature* 428, 557–560.

Webster, M.A., Kaping, D., Mizokami, Y., & Duhamel, P. (2004). Adaptation to natural facial categories. *Nature,* 428(6982), 557-561.

Webster, M.A., Werner, J.S., Field, D.J., Clifford, C.W.G., & Rhodes, G. (eds). (2005). Adaptation and the phenomenology of perception. Fitting the mind to the world: Adaptation and aftereffects in high level vision. *Oxford University Press , Oxford, UK,* 241-277.

Wichmann, F.A., & Hill, N.J. (2001a) The psychometric function: I. Fitting, sampling, and goodness of fit. *Perception & Psychophysics* 63,(8), 1293-1313.

Williams, L. M., Senior, C., David, A. S., Loughland, C. M., & Gordon, E. (2001). In search of the 'Duchenne smile': Evidence from eye movements. *Journal of Psychophysiology,* 15(2), 122-127.

Wilson, H.R., Loffler, G., & Wilkinson, F. (2002). Synthetic faces, face cubes, and the geometry of face space, *Vision Research* 42, 2909-2923.

Wilson, J. R., and Vandenberg, S. G. (1978). Sex differences in cognition: Evidence from the Hawaii Family Study. In McGill, T. E., Dewsbury, D. A., and Sachs, B. D. (eds.) In: *Sex and behavior. New York: Plenum.*

Wright, D.B., & Sladden, B. (2003). An own gender bias and the importance of hair in face recognition. *Acta Psychologica* 114(1), 101-114.

Yamaguchi, M. K., Hirukawa, T., & Kanazawa, S. (1995). Judgment of gender through facial parts. *Perception,* 24(5), 563-575.

Yang, H., Shen, J., Chen, J., & Fang, F. (2010). Face adaptation improves gender discrimination. *Vision Research,* in press.

Yarbus, A. L. (1967). Eye movements and vision. New York: Plenum Press.

Yin, R.K (1969). Looking at upside-down faces. *Journal of Experimental Psychology* 81(1), 141.

Yip, A.W., & Sinha, P. (2002). Contribution of color to face recognition. *Perception*

31(8), 995-1003.

Young, A.W., Hellawell, D., & Hay, D.C. (1987). Configurational information in face perception. *Perception* 16(6), 747-759.

Zebrowitz, L.A., Rhodes, G. (2004) Sensitivity to "bad genes" and the anomalous face overgeneralization effect: Cue validity, cue utilization, and accuracy in judging intelligence and health. *Journal of Nonverbal Behavior* 28(3), 167-185.